Advances in Mobile Cloud Computing Systems

OTHER COMMUNICATIONS BOOKS FROM CRC PRESS

Analytical Evaluation of Nonlinear Distortion Effects on Multicarrier Signals
Theresa Araújo
ISBN 978-1-4822-1594-6

Architecting Software Intensive Systems: A Practitioners Guide
Anthony J. Lattanze
ISBN 978-1-4200-4569-7

Cognitive Radio Networks: Efficient Resource Allocation in Cooperative Sensing, Cellular Communications, High-Speed Vehicles, and Smart Grid
Tao Jiang, Zhiqiang Wang, and Yang Cao
ISBN 978-1-4987-2113-4

Complex Networks: An Algorithmic Perspective
Kayhan Erciyes
ISBN 978-1-4665-7166-2

Data Privacy for the Smart Grid
Rebecca Herold and Christine Hertzog
ISBN 978-1-4665-7337-6

Generic and Energy-Efficient Context-Aware Mobile Sensing
Ozgur Yurur and Chi Harold Liu
ISBN 978-1-4987-0010-8

Just Ordinary Robots: Automation from Love to War
Lamber Royakkers and Rinie van Est
ISBN 978-1-4822-6014-4

Machine-to-Machine Communications: Architectures, Technology, Standards, and Applications
Vojislav B. Misic and Jelena Misic
ISBN 978-1-4665-6123-6

Managing the PSTN Transformation: A Blueprint for a Successful Migration to IP-Based Networks
Sandra Dornheim
ISBN 978-1-4987-0103-7

MIMO Processing for 4G and Beyond: Fundamentals and Evolution
Edited by Mário Marques da Silva and Francisco A. Monteiro
ISBN 978-1-4665-9807-2

Mobile Evolution: Insights on Connectivity and Service
Sebastian Thalanany
ISBN 978-1-4822-2480-1

Network Innovation through OpenFlow and SDN: Principles and Design
Edited by Fei Hu
ISBN 978-1-4665-7209-6

Neural Networks for Applied Sciences and Engineering: From Fundamentals to Complex Pattern Recognition
Sandhya Samarasinghe
ISBN 978-0-8493-3375-0

Rare Earth Materials: Properties and Applications
A.R. Jha
ISBN 978-1-4665-6402-2

Requirements Engineering for Software and Systems, Second Edition
Phillip A. Laplante
ISBN 978-1-4665-6081-9

Security for Multihop Wireless Networks
Edited by Shafiullah Khan and Jaime Lloret Mauri
ISBN 978-1-4665-7803-6

The Future of Wireless Networks: Architectures, Protocols, and Services
Edited by Mohesen Guizani, Hsiao-Hwa Chen, and Chonggang Wang
ISBN 978-1-4822-2094-0

The Internet of Things in the Cloud: A Middleware Perspective
Honbo Zhou
ISBN 978-1-4398-9299-2

The State of the Art in Intrusion Prevention and Detection
Al-Sakib Khan Pathan
ISBN 978-1-4822-0351-6

ZigBee® Network Protocols and Applications
Edited by Chonggang Wang, Tao Jiang, and Qian Zhang
ISBN 978-1-4398-1601-1

TO ORDER
Call: 1-800-272-7737 • Fax: 1-800-374-3401 • E-mail: orders@crcpress.com

Advances in Mobile Cloud Computing Systems

Edited by
F. Richard Yu
Victor C.M. Leung

CRC Press is an imprint of the
Taylor & Francis Group, an **informa** business

MATLAB® and Simulink® are trademarks of The MathWorks, Inc. and are used with permission. The Math-Works does not warrant the accuracy of the text or exercises in this book. This book's use or discussion of MATLAB® and Simulink® software or related products does not constitute endorsement or sponsorship by The MathWorks of a particular pedagogical approach or particular use of the MATLAB® and Simulink® software.

CRC Press
Taylor & Francis Group
6000 Broken Sound Parkway NW, Suite 300
Boca Raton, FL 33487-2742

© 2016 by Taylor & Francis Group, LLC
CRC Press is an imprint of Taylor & Francis Group, an Informa business

No claim to original U.S. Government works

Printed on acid-free paper
Version Date: 20151015

International Standard Book Number-13: 978-1-4987-1509-6 (Hardback)

Library of Congress Cataloging-in-Publication Data

Advances in mobile cloud computing systems / edited by F. Richard Yu and Victor C.M. Leung.
    pages cm
Includes bibliographical references and index.
ISBN 978-1-4987-1509-6 (acid-free paper) 1. Cloud computing. 2. Mobile computing.
I. Yu, F. Richard, editor. II. Leung, Victor Chung Ming, 1955- editor.

QA76.585.A384 2016
004.67'82--dc23   2015024986

Visit the Taylor & Francis Web site at
http://www.taylorandfrancis.com

and the CRC Press Web site at
http://www.crcpress.com

Contents

Preface

This is a brief journey through *Advances in Mobile Cloud Computing Systems*.

Introduction

There is a phenomenal burst of research activities in mobile cloud computing systems, which extends cloud computing functions, services, and results to the world of future mobile communications applications, and the paradigm of cloud computing and virtualization to mobile networks. Mobile applications demand greater resources and improved interactivity for better user experience. Resources in cloud computing platforms such as Amazon, Google AppEngine, and Microsoft Azure are a natural fit to remedy the lack of local resources in mobile devices. On the other hand, wireless network virtualization enables abstraction and sharing of infrastructure and radio spectrum resources, the overall expenses of wireless network deployment and operation can be reduced significantly. The availability of cloud computing resources on a pay-as-you-go basis, the advances in wireless network virtualization, software-defined networking (SDN), device-to-device (D2D) communications, local cloudlets, and the emergence of advanced wireless networks such as cloud-based radio access networks (C-RANs) create a new space of rich research problems.

Mobile cloud computing systems have opened up several areas of research, which have been explored extensively and continue to attract research and development. This book essentially features some of the major advances in the research on mobile cloud computing systems. The chapters in this book, contributed by leading experts in the field, cover different aspects of modeling, analysis, design, and optimization of algorithms, protocols, and architectures of mobile cloud computing systems. A summary of all of the chapters is provided in the following sections.

Chapter 1, authored by Zhiyuan Yin, F. Richard Yu, and Shengrong Bu, considers a mobile cloud computing system with

vii

a telecom operator cloud, in which cloud mobile media (CMM) service providers and the size of the CMM service can be dynamically selected. In this way, a telecom operator can strengthen its relationships with end-users and third-party CMM service providers by acting as a service and billing aggregator. The authors formulate the problem of determining the CMM service price decision, resource allocation of wireless network resources, and interference management as a multilevel Stackelberg game. In addition, they propose an iteration algorithm to obtain the Stackelberg equilibrium solution.

Chapter 2, authored by Yonggang Wen, Weiwen Zhang, and Kyle Guan, focuses on the energy-efficient task execution policy in mobile cloud computing. The objective is to reduce energy consumption on both mobile devices and the cloud for executing transcoding tasks. For the mobile device, the authors find an operational region to determine whether the task should be off-loaded or not. For the cloud, they leverage a Lyapunov optimization framework and propose an online algorithm to reduce energy consumption on service engines while achieving queue stability.

Chapter 3, authored by Hyunseok Chang, Adiseshu Hari, Sarit Mukherjee, and T.V. Lakshman, presents such a hybrid cloud architecture, called a *Proximity Cloud*, which is designed to deliver low-latency, bandwidth-efficient, and resilient end-user services with a global footprint. The following aspects of the Proximity Cloud are described: a new paradigm for delivering services to the edge that combines the advantages of data-center–based service delivery with local presence at the edge, an instantiation of this paradigm on the OpenStack cloud management platform with OpenStack extensions to support NAT-friendly and secure virtual tenant networks, and the quantitative evaluation of the Proximity Cloud and two edge apps that we deploy in the Proximity Cloud: 3D indoor localization and video surveillance. The authors also describe a distributed Hadoop implementation in the Proximity Cloud.

Chapter 4, authored by Mohamad Kalil, Khalim Amjad Meerja, Ahmed Refaey, and Abdallah Shami, describes RAN network virtualization in clouds. The sharing of resources of virtualized 5G wireless broadband networks simultaneously by multiple mobile network operators (MNOs) or simply called service providers (SPs) is discussed. Then the system model for sharing of resources using wireless resource virtualization framework is provided. The authors also discuss binary integer programming (BIP) formulation of the resource sharing problem.

Chapter 5, authored by Haleh Khojasteh, Jelena Mišić, and Vojislav B. Mišić, presents a simple yet effective admission control mechanism that can be easily added to an existing cloud center and investigate its performance as the function of system load and baseline partitioning of servers into pools. The authors discuss the system model of a cloud data center with the pooling mechanism and present its performance. They also present the proposed admission control mechanism through two algorithms of increasing complexity and its performance improvements.

Chapter 6, authored by Yujin Li and Wenye Wang, investigates the fundamental questions in mobile cloudlet spectrum: What is the computing performance of a mobile cloudlet under intermittent device connectivity, and if and under what conditions can mobile cloudlet support mobile applications? Based on cloudlet properties, the authors also derive upper and lower bounds on the computing capacity and long-term computing speed of a mobile cloudlet. An initiator can use these bounds to decide whether to upload a task to remote clouds or utilize nearby mobile cloudlets.

Chapter 7, authored by Mazhar Ali, Muhammad Usman Shahid Khan, Assad Abbas, and Samee U. Khan, proposes a software piracy control framework for mobile cloud applications. The proposed framework considers the piracy control issue as one of the offshoots of access control. The software license ensures that the software is executed only by the party that is authorized or has been granted access to execute. In the same manner, the proposed framework gives access to the requesting user to execute the application on the cloud only if the user is authorized to do so. The license verification and access grant in the proposed framework is ticket based, where the credentials and parameters are verified by the possession of the valid ticket or otherwise. The ticket is issued for a specified period and the execution of the application after that time will require the acquisition of the new ticket.

Chapter 8, authored by Yegui Cai, F. Richard Yu, and Shengrong Bu, considers how to dynamically configure C-RAN to enhance MCC services' performance in a holistic framework. In this work, they optimize the end-to-end TCP throughput performance of MCC users in next-generation cellular networks. They investigate the trade-off between the systematic efficiency and the fairness among MCC users.

A parameter is introduced to study such trade-off taking into account the delayed CSI in C-RAN. With this parameter, they reformulate the problem to maximize the Jain's fairness index in MCC systems.

Chapter 9, authored by Heli Zhang, Weidong Wang, Xi Li, and Hong Ji, studies the user association issue in the cloud-RAN SCN scenario. The authors establish a user association optimization problem with the aim of minimizing network latency. The latency is deduced by a minimal potential delay fairness function (MPDF). Moreover, considering the rising CO_2 emission and to guarantee the network throughput, energy saving and interference limitation are also incorporated in the optimization. To solve the user association optimization problem, a three-phrase search algorithm (TPSA) is proposed using the concept of Pareto optimality. Under the help of TPSA, appropriate small cells and physical resources are chosen for users while minimizing overall energy consumption and reducing network interference.

Conclusion

This summary should help the reader follow the rest of this book. These chapters essentially feature some of the major advances in the research on mobile cloud computing systems. Therefore, the book will be useful to both researchers and practitioners in this area. The reader will find the rich set of references in each chapter particularly valuable.

F. Richard Yu
Victor C.M. Leung

MATLAB® is a registered trademark of The MathWorks, Inc. For product information, please contact:

The MathWorks, Inc.
3 Apple Hill Drive
Natick, MA 01760-2098 USA
Tel: 508-647-7000
Fax: 508-647-7001
E-mail: info@mathworks.com
Web: www.mathworks.com

Editors

F. Richard Yu received his PhD in electrical engineering from the University of British Columbia (UBC), Vancouver, British Columbia, Canada in 2003. From 2002 to 2004, he was with Ericsson, in Lund, Sweden, where he worked on the research and development of wireless mobile systems. From 2005 to 2006, he was with a start-up company in California, where he worked on research and development in the areas of advanced wireless communication technologies and new standards. He joined the Carleton School of Information Technology and the Department of Systems and Computer Engineering at Carleton University in 2007, where he is currently an associate professor. He received the IEEE Outstanding Leadership Award in 2013, Carleton Research Achievement Award in 2012, the Ontario Early Researcher Award (formerly Premier's Research Excellence Award) in 2011, the Excellent Contribution Award at IEEE/IFIP TrustCom 2010, the Leadership Opportunity Fund Award from the Canada Foundation of Innovation in 2009, and the Best Paper Award at the IEEE ICC 2014, Globecom 2012, IEEE/IFIP TrustCom 2009, and International Conference on Networking 2005. His research interests include cross-layer/cross-system design, security, green IT, and QoS provisioning in wireless-based systems.

He serves on the editorial boards of several journals, including as co-editor-in-chief for *Ad Hoc & Sensor Wireless Networks*, lead series editor for *IEEE Transactions on Vehicular Technology*, *IEEE Communications Surveys & Tutorials*, *EURASIP Journal on Wireless Communication and Networking*, Wiley's *Security and Communication Networks*, and *International Journal of Wireless Communications and Networking*; he is a guest editor for *IEEE Transactions on Emerging Topics in Computing* and a special issue of *Advances in Mobile Cloud Computing*, and a guest editor of the *IEEE Systems Journal* special issue on smart grid communications systems. He has served on the technical program committee (TPC) of numerous conferences, as the TPC co-chair of IEEE GreenCom'14, INFOCOM-MCV'15, Globecom'14, WiVEC'14, INFOCOM-MCC'14,

Globecom'13, GreenCom'13, CCNC'13, INFOCOM-CCSES'12, ICC-GCN'12, VTC'12S, Globecom'11, INFOCOM-GCN'11, INFOCOM-CWCN'10, IEEE IWCMC'09, VTC'08F, and WiN-ITS'07; as the publication chair of ICST QShine'10; and the co-chair of ICUMT-CWCN'09. Dr. Yu is a registered professional engineer in the province of Ontario, Canada.

Victor C.M. Leung is a professor of electrical and computer engineering and holds the TELUS Mobility Research Chair at the University of British Columbia (UBC), Vancouver, British Columbia, Canada. His research is in the areas of wireless networks and mobile systems, where he has coauthored more than 800 technical papers in archival journals and refereed conference proceedings, several of which have won best paper awards. Dr. Leung is a fellow of the IEEE, the Royal Society of Canada, the Canadian Academy of Engineering, and the Engineering Institute of Canada. He is serving/has served on the editorial boards of *IEEE Journal on Selected Areas in Communications, IEEE Transactions on Computers, IEEE Transactions on Wireless Communications,* and *IEEE Transactions on Vehicular Technology, and IEEE Transactions on Wireless Communications Letters,* and several other journals. He has provided leadership to the technical committees and organizing committees of numerous international conferences. Dr. Leung is a recipient of an APEBC Gold Medal, NSERC Postgraduate Scholarships from 1977 to 1981, a 2012 UBC Killam Research Prize, and an IEEE Vancouver Section Centennial Award.

Contributors

Assad Abbas
North Dakota State University
Fargo, North Dakota

Mazhar Ali
North Dakota State University
Fargo, North Dakota

Shengrong Bu
University of Glasgow
Glasgow, United Kingdom

Yegui Cai
Carleton University
Ottawa, Ontario, Canada

Hyunseok Chang
Bell Labs
Alcatel-Lucent
New Providence, Jew Jersey

Kyle Guan
Bell Labs
Alcatel-Lucent
New Providence, Jew Jersey

Adiseshu Hari
Bell Labs
Alcatel-Lucent
New Providence, Jew Jersey

Hong Ji
Beijing University of Posts and
    Telecommunications
Beijing, People's
    Republic of China

Mohamad Kalil
Western University
London, Ontario, Canada

**Muhammad Usman Shahid
Khan**
North Dakota State University
Fargo, North Dakota

Samee U. Khan
North Dakota State University
Fargo, North Dakota

Haleh Khojasteh
Department of Computer Science
Ryerson University
Toronto, Ontario, Canada

T.V. Lakshman
Bell Labs
Alcatel-Lucent
New Providence, Jew Jersey

Xi Li
Beijing University of Posts and
    Telecommunications
Beijing, People's
    Republic of China

Yujin Li
Department of Electrical and
    Computer Engineering
North Carolina State University
Raleigh, North Carolina

Khalim Amjad Meerja
Western University
London, Ontario, Canada

Jelena Mišić
Department of Computer Science
Ryerson University
Toronto, Ontario, Canada

Vojislav B. Mišić
Department of Computer Science
Ryerson University
Toronto, Ontario, Canada

Sarit Mukherjee
Bell Labs
Alcatel-Lucent
New Providence, Jew Jersey

Ahmed Refaey
Mircom Technologies Ltd.
Toronto, Ontario, Canada

Abdallah Shami
Western University
London, Ontario, Canada

Weidong Wang
Beijing University of Posts and
    Telecommunications
Beijing, People's
    Republic of China

Wenye Wang
Department of Electrical and
    Computer Engineering
North Carolina State University
Raleigh, North Carolina

Yonggang Wen
Nanyang Technological
    University
Singapore, Singapore

Zhiyuan Yin
Carleton University
Ottawa, Ontario, Canada

F. Richard Yu
Carleton University
Ottawa, Ontario, Canada

Heli Zhang
Beijing University of Posts and
    Telecommunications
Beijing, People's
    Republic of China

Weiwen Zhang
Nanyang Technological
    University
Singapore, Singapore

CHAPTER 1

Mobile Cloud Computing with Telecom Operator Cloud

Zhiyuan Yin, F. Richard Yu, and Shengrong Bu

CONTENTS

1

1.1   Introduction

Recently, cloud computing has drawn a lot of attention from both academia and industry. It has advantages over traditional computing paradigms, such as avoiding capital investments and operational expenses for end-users. The essential characteristics of cloud computing include on-demand self-service, broadband network access, resource pooling, rapid elasticity, and measured service [14]. Furthermore, with recent advances in mobile communication technologies, mobile devices, and mobile applications, more and more end-users access cloud computing systems via mobile devices, such as smartphones. As such, mobile cloud computing is widely considered as a promising computing paradigm with a huge market. In traditional mobile computing systems, mobile devices usually have limited computing and storage capabilities. In contrast, utilizing the powerful computing and storage resources available in the cloud environments, mobile cloud computing can enable the use of cutting-edge multimedia services. In the cloud, the resources have much higher processing and storage capacities compared to what traditional mobile devices can provide. And thus, the cloud can offer a much richer media experience than current mobile applications.

There are many promising *cloud mobile media* (CMM) services based on mobile cloud computing, including media storage and downloading services, audio services, and interactive services (e.g., multi-way video conferencing, advertisements, and mobile multi-player gaming). CMM services will not only make the end-users enjoy a richer media experience from a mobile device, but a more important aspect is that CMM services will also offer new opportunities for CMM *service providers* (SPs) and telecom operators to offer end-users rich media services that can be delivered efficiently such that end-users are satisfied and have a good quality of experience. For telecom operators, CMM services will narrow the growing gap between the growth in data usage by end-users and revenue earned by SPs. Despite the potential of CMM services, several research challenges still need to be addressed. These include the following: the availability and accessibility of media services, mobile data integrity, user privacy, energy efficiency, response time, and the *quality of service* (QoS) over wireless networks. Among these challenges in deploying CMM services, one of the most important is the response time experienced by end-users, which is highly dependent on the quality and speed of the network. Indeed, networking has become a bottleneck, and thus, it has a significant impact on the quality of cloud services. This problem becomes more severe in CMM services due to the scarce network capacity, higher bit error rate, and user mobility in wireless networks. One of the main reasons behind the aforementioned problem is that cloud computing and network communications are not jointly designed and optimized. Therefore, when the end-users use CMM services they may suffer long response times and a degraded QoS. In this chapter, we propose an approach based on game theory to consider the joint design of cloud computing and wireless networks. In addition, we introduce the deployment of the *telecom operator cloud* (TOC) as a promising approach for "interchanging and mixing" the CMM services from different third-party CMM SPs.

Most current CMM services are offered by *over-the-top* (OTT) players, which provide easy-to-adopt, on-demand services by exploiting ubiquitous connectivity [27]. OTT SPs can offer tons of CMM services, which contain media storage services, downloading services, and interactive services (e.g., multi-way video conferencing, and cloud gaming) [7,28]. However, these OTT players do not provide communication or connectivity services, which are offered by traditional telecom operators. Thus, it is necessary to provide a quantitative approach to jointly consider the following problems in a mobile cloud computing environment: the price to charge end-users for CMM services, resource allocation, and the interference management of wireless networks.

To solve the mentioned problems met in the mobile cloud computing environments, an approach based on game theory is proposed to model the interactions between CMM SPs and the end-users in the *heterogeneous wireless networks* (HWNs), where the objective is to benefit both parties.

In this chapter, we jointly study the operations of cloud computing and wireless networks in a mobile computing environment with a TOC. The objective is to improve CMM services. The distinct features of this work are outlined in the following:

- We consider a mobile cloud computing system with TOC, in which CMM SPs and the size of the CMM service can be dynamically selected. In this way, a telecom operator can strengthen its relationships with end-users and third-party CMM service providers by acting as a service and billing aggregator. This can be accomplished using network virtualization technology [14].

- The operations of wireless networks (e.g., resource allocation and interference management) are optimized according to the following: CMM traffic, real-time price provided by the CMM SPs, and the cost associated with the CMM services.

- We formulate the problem of determining the CMM service price decision, resource allocation of wireless networks, and interference management as a multi-level Stackelberg game. Stackelberg game has been successfully used in cooperative communication networks, and other areas [41]. To analyze the proposed game, we use a backward induction method [20] that can capture the sequential dependencies of the decisions in the stages of the game model. In addition, we propose an iteration algorithm to obtain the Stackelberg equilibrium solution.

- Extensive simulations are performed to investigate the performance of the proposed techniques. Based on the simulation results, we verify the convergence of the Stackelberg equilibrium iteration algorithms.

- New challenges arise, which have not been addressed by existing research, when jointly considering the dynamics of cloud, CMM services, and wireless networks. We believe that the research and results that are presented will open a new avenue and motive additional research for considering the operations of cloud and wireless networks in a mobile cloud computing environment.

1.2 Background and Related Works

This section provides a survey of literature related to this chapter.

Several definitions of mobile cloud computing are available. In [35], the authors define the mobile cloud computing as "a rich mobile computing technology that leverages unified elastic resources of varied clouds and network technologies toward unrestricted functionality, storage, and mobility." It serves a multitude of mobile devices anywhere, anytime through the channel of ethernet of the Internet regardless of heterogeneous environments and platforms based on the "pay-as-you-use" principle.

Another definition of mobile cloud computing is described in [15]:

> commonly, the term mobile cloud computing means to run an application such as Google's Gmail for Mobile on a remote resource rich server, while the mobile device acts like a thin client connection over to a remote server through wireless network. Some other examples of this type are Facebook's location-aware services, Twitter for mobile, mobile weather widgets, etc.

The authors also consider mobile devices themselves as resource providers to set up a mobile P2P network. Thus, this approach supports user mobility, and recognizes the potential of mobile clouds to perform collective sensing as well.

In [23], the authors describe that "mobile cloud computing often involves three foundations, namely cloud computing, mobile computing and networking, and can be considered as an emerging cloud service model following the trend to extend the cloud to the edge of networks."

In this chapter, we adopt the definition made by [8]:

> mobile cloud computing is an emergent mobile cloud paradigm which leverages mobile computing, networking, and cloud computing to study mobile service models, develop mobile cloud infrastructures, platforms, and service applications for mobile clients. Its primary object is to deliver location-aware mobile services with mobility to users based on scalable mobile cloud resources in networks, computers, storages, and mobile devices. Its goal is to deliver them with secure mobile cloud resources, service applications, and data using energy-efficient mobile cloud resources in a "pay-as-you-use" model.

Although some excellent works have been done to study cloud computing and wireless networks [4,6,8,29], these two important areas have traditionally been addressed separately in the literatures. However, from the perspective of end-to-end applications [13], both cloud and wireless networks are part of the entire system. The experience in end-to-end applications (e.g., video and TCP-based applications) indicates that the optimized performance in one segment of the whole system does not guarantee the end-to-end performance [3,10].

To ensure the optimal usage of the resources of clouds and wireless networks and enable the scalability of the CMM service users [25], joint cloud and wireless networks operations should be used for each CMM service client [45].

1.2.1   Features of Mobile Cloud Computing

The primary features of mobile cloud computing are shown in Figure 1.1 and are described in detail here:

- *Resource Management*: Mobile clouds can manage the resources more freely and enable resources provisioning and de-provisioning automatically. The resources that need to be managed computing resources, network resources, as well as the resources of mobile devices.

- *Security and Privacy*: This feature relies on the body of the security capabilities, technologies, processes, and practices. It is designed

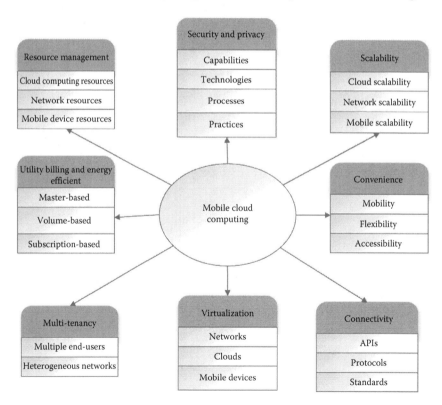

Figure 1.1   The primary features of mobile cloud computing.

to protect mobile devices, heterogeneous networks, clouds, and the data from attacks, damage, and illegal access.

- *Scalability*: This feature considers three dimensions for scalability in the mobile cloud computing environments that include cloud scalability, network scalability, and the scalability of the mobile devices.

- *Convenience*: In the mobile cloud computing environments, the end-users can access the cloud resources (services and applications) at anytime from anywhere.

- *Connectivity*: This feature adopts the well-designed APIs, protocols, and standards offered by existing work on mobile clouds to enable easy and secure connectivity between different networks, and third-party applications or systems.

- *Virtualization*: There are three types of virtualization (network virtualization, cloud virtualization, and virtualization mobile devices) that can be used in the mobile cloud computing environments.

- *Multi-tenancy*: This feature allows the mobile cloud to support multiple end-users in heterogeneous networks.

- *Utility Billing and Energy Efficient*: The mobile cloud provides several service billing models, such as meter-based, volume-based, and subscription-based.

1.2.2   Mobile Cloud Computing Business Service Models

Similar to the cloud computing model, for achieving low-cost media services by using the "pay-as-you-use" approach, mobile cloud computing can also adopt the utility billing model to acquire resources and provide the media services. There are several business service models based on mobile cloud computing as listed in the following:

- *Mobile Software-as-a-Service* (*MSaaS*): The mobile cloud computing can enable *Software-as-a-Service* (SaaS) and its related functions, which are provided to the end-users with additional features, such as mobility, location-awareness, and accessibility from anywhere at anytime. Through this model, the end-users can access the mobile application services deployed on the cloud using the wireless communication technology.

- *Mobile Platform-as-a-Service (MPaaS)*: The mobile cloud computing delivers mobile platforms as a cloud service. It integrates the
  mobile application management, mobile operator management,
  and mobile device management built on the cloud to provide maximum scalability and security.

- *Mobile Infrastructure-as-a-Service (MIaaS)*: The cloud infrastructure and its resources are provided to the end-users using "pay-as-
  you-use" approach. Through MIaaS, all the computing and storage
  capability of the cloud can be provisioned, manage, and returned
  according to the demand of the mobile end-users' requests.

- *Mobile Network-as-a-Service (MNaaS)*: The infrastructure of heterogeneous networks and its related resources are provided by a vendor or "broker" to the mobile end-users in response to on-demand
  requests. This allows a desirable wireless network infrastructure to
  be dynamically configured, deployed, and structured for the mobile
  connectivity to an existing cloud infrastructure. MNaaS provides
  mobile networking infrastructures as a service [13]. The primary
  advantage of MNaaS is higher scalability and elasticity. In addition, MNaaS requires a relatively low start-up cost for a network
  service vendor or "broker." An infrastructure provider or independent telecom operator builds and operates a network (using wireless or transport connectivity) and sells its communication access
  capabilities to the third parties of clouds charging a price according to capacity utilized by the end-user [32]. An example of MNaaS
  is OpenStack that is an open-source cloud operating system [12].
  It allows end-users to create their own networks, control traffic,
  and connect servers and devices to one or more networks.

- *Mobile App-as-a-Service (MAaaS)*: It refers to a service business model where diverse mobile applications can be deployed,
  managed, hosted, and monitored.

- *Mobile Testing-as-a-Service (MTaaS)*: It refers to a service business
  model where various mobile-based testing devices, tools, and services are provided by a vendor or "brokering" as resources to its
  mobile clients. Clients use this service to help support the testing
  of their mobile-based softwares and applications, and are charged
  using a "pay-as-you-use" or other business models.

- *Mobile Community-as-a-Service (MCaaS)*: It refers to a service business model where various mobile social networks and communities can be dynamically established and managed to provide social community services. MCaaS can also be used to provide networking to mobile end-users using a "pay-as-you-use" or other business models.

- *Mobile Multimedia-as-a-Service (MMaaS)*: It refers to a service business model where rich media services based on the application services (e.g., high-quality movies and digital games) are deployed, managed, and hosted to deliver the media service to the mobile end-users using a "pay-as-you-use" or other business models.

1.2.3   Network Virtualization

Network virtualization is a game-changing technology that enables operators to meet all these evolving demands of its users giving the operators the ability to scale network capacity dynamically and adopt a range of innovative business models.

Virtualization essentially decouples the network hardware from software by introducing a virtual resource manager layer that mediates between the network elements and the software-based network controllers. *Software-defined networking* (SDN) is an emerging network architecture that is considered as one of the most promising technologies for realizing virtual networks, especially for managing the control of the network [33,38]. The four key features of SDN are summarized in the following:

- Separation of the control plane from the data plane

- A centralized controller and view of the network

- Open interfaces between the devices in the control plane and those in the data plane

- Programmability of the network by external applications

Similar to the wired network virtualization discussed earlier, wireless network virtualization can have a very broad scope ranging from spectrum sharing, infrastructure virtualization, to air interface virtualization. The physical infrastructures owned by one or more telecom operators can be shared among multiple service providers. Wireless network virtualization needs the physical wireless infrastructure and radio resources to be abstracted and isolated to a number of virtual resources, which can be acquired by different SPs. In other words, virtualization of networks regardless of wired or wireless, can be considered as the process of splitting the entire network infrastructure into virtual resources [47]. However, the distinctive properties of the wireless environments, in terms of time-various channels, attenuation, mobility, broadcast, etc., make wireless network virtualization more complicated.

In this chapter, we define *wireless network virtualization* (WNV) as the technology in which infrastructure resources and physical radio resources can be abstracted and sliced into virtual wireless network resources that have certain corresponding functionalities, and can be shared by multiple parties through isolation from each other. In other words, virtualizing a wireless network is to realize the process of abstracting, slicing, isolating, and sharing the wireless networks. Above all, it is the key concept of the MNaaS business model. Since wireless network resources are sliced into multiple slices, the terms virtual slice and virtual network have a similar meanings to the term virtual wireless network resources. To avoid confusion, we use the term network virtualization to refer to this entity.

1.2.4 Mobile Cloud Service General Model

According to [26], the total revenues of entertainment mobile cloud services (e.g., Amazon's Cloud Drive and Apple's iCloud) are expected to be up to $39 billion by 2016. This situation leads to the explosive growth in mobile data usage, and also means that there will be a dramatic shift regarding how mobility is being used in entertainment areas. The limitations of mobile cloud services are summarized in the following:

- Limited scalability in network bandwidths and traffic support

- Limited probability and connectivity between wireless networks that are owned by different telecom operators

- Limited resource sharing and insufficient usage of network resources

- Less green computing in networks

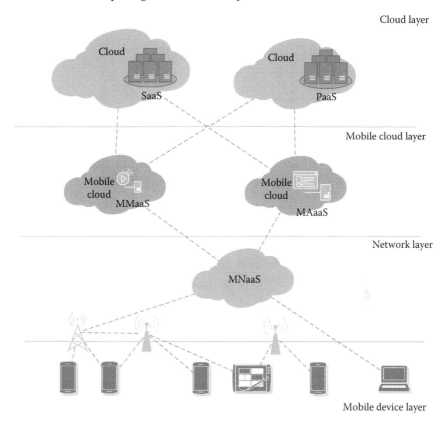

Figure 1.2   A general model for mobile cloud services.

The issues discussed earlier have become the key challenges of mobile cloud computing. To achieve elasticity and scalability for mobile networking and communication services, the architecture of mobile service system is deployed using the following layers as shown in Figure 1.2:

- *Cloud layer*: Scalable cloud infrastructures and platforms, where diverse back-end mobile application servers can be executed, managed, and maintained on the selected cloud for deploying a specific service (e.g., PaaS and SaaS). The objective is to share the computing resources, improve the system utilization, and reduce the cost of the utility billings.

- *Mobile cloud layer*: This layer consists of the essential services and features critical to mobile cloud computing, such as energy-efficient, provisioning and management of mobile cloud resources, service for mobile application brokering, mobile multimedia services, mobile application services, and mobile cloud security management, etc.

- *Network layer*: This layer groups, manages, and delivers the network resources via a MNaaS business model. In addition, the network virtualization and energy-saving techniques are used to achieve the elasticity and scalability of the network resources, which helps achieve higher network resource utilization.

- *Mobile device layer*: This layer provides the mobile connectivity interfaces to the mobile end-users so that the users can have a consistent and comprehensive access to mobile cloud application services. The essential features of this layer include secure end-to-end mobile transactions and connectivity, as well as maintaining the privacy of end-users.

The mobile service general model offers many advantages as described here:

- The model increases the elasticity and scalability of wireless networks, as well as provide better resource utilization and sharing of resources.

- The model reduces the capital expenditures for the development of diverse mobile network resources, which support a variety of access technologies and combine different market segments.

- The model reduces the cost of operations via use of energy-efficiency techniques and technologies, as well as improves the overall utilization of the network.

- Due to the unified access, this model can reduce the development cost as well.

1.2.5 Heterogeneous Wireless Networks

HWNs approach is introduced to solve the "last mile" issue in the mobile cloud computing environments. The deployment of HWNs with small cells is an important technique to increase the energy efficiency of wireless cellular networks [34]. Currently, more and more mobile data traffic is being generated by mobile end-users because of

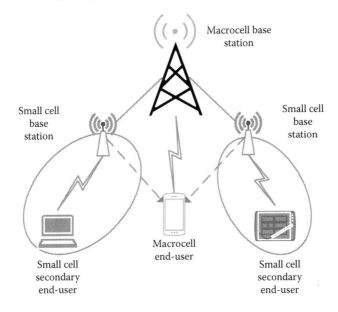

Macrocell base station

Small cell base station

Small cell base station

Macrocell end-user

Small cell secondary end-user

Small cell secondary end-user

Figure 1.3   The heterogenous wireless networks model.

activities, such as high-quality movies, and playing the digital video games. In HWNs, macrocells have been deployed to provide a larger coverage area, and to improve the mobility of the devices in cellular networks. However, a disadvantage of macrocells is that they cannot deliver high data rates at indoor environments. To solve this issue, small cells are deployed into HWNs to provide higher data rates and to enhance the in-building coverage as well. Due to the coverage area of small cells, they require much less transmission power than the macrocells. Thus by using small cells, base stations become more energy-efficient when providing broad coverage. However, a large number of small cells may cause interference among the macrocells and other small cells, which leads to degradation of performance and energy efficiency of the whole network [17]. Therefore, a deployment of base stations with different cell sizes (macro and small) is desirable in energy-efficient networks. The HWNs model is shown in Figure 1.3.

1.3   Dynamic Cloud and Wireless Network Operations in Mobile Cloud Computing Environments

In this chapter, we first describe a mobile cloud computing system including several third-party CMM SPs, which offer their services to a telecom operator and use the telecom operator as a "broker." Then we introduce HWNs with the small cells model to improve the network performance in terms of capacity and energy efficiency. Finally, we formulate the whole system as a three-stage Stackelberg game and the simulation results are discussed at the end of this chapter.

1.3.1   System Description and Formulation

As shown in Figure 1.4, we consider a mobile cloud computing system with several third-party CMM SPs. The system has a telecom operator cloud that can "mix and interchange" resources offered by different third-party CMM SPs, and HWNs contain both macro and small cell base stations. In addition, the problem of joint cloud and wireless networks operations in this system is formulated as a three-stage Stackelberg game.

1.3.1.1   Telecom Operator Cloud and Third-Party CMM SPs

Future media service will definitely be provided by clouds. The multimedia service may come from different cloud service providers, different network types, different technologies, etc. In this chapter, we mainly consider the CMM service model provided by different third-party CMM SPs. Different CMM SPs offer rich multi-media services to the end-users, including the streaming media, interactive service, and mobile gaming, etc. In the previous section, we have discussed that most of the CMM SPs will choose to partner with the telecom operators rather than to pay. In this scenario, telecom operators will pool variety of third-party CMM SPs and offer a virtually unlimited selection of customized and diverse services for end-users.

Telecom operators are making a stronger push to monetize data traffic through OTT-style services and applications. They also attempt to restrict these traffic-heavy, low-revenue OTT services such as streaming media, but this may have negative impacts on user experience, even violates the regulations sometimes.

Operators can play a natural role in the cloud computing, providing the reliable and low-cost connectivity service for any other third-party clouds, but this is just an initial step. Some of the pioneers in this area have already explored a new TOC model [5,14,16]. On one hand, telecom operators can use the powerful storage and computing

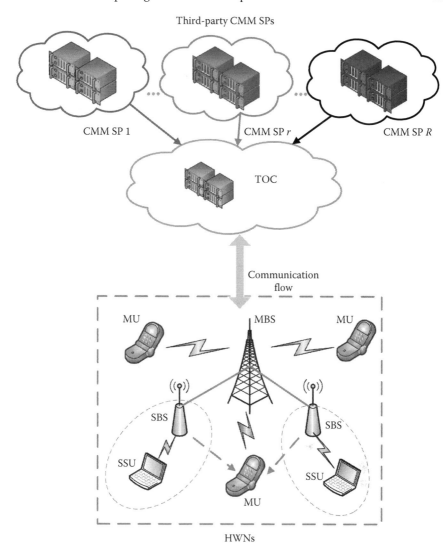

Figure 1.4 A mobile cloud computing system with telecom operator cloud. (From Yin, Z. et al., *IEEE Trans. Wireless Commun.*, 14, 4020, 2015. With permission.)

capabilities offered by the cloud for network management, such as billing, etc. In this case, telecom operators are cloud users. On the other hand, telecom operators can also be cloud providers as well. For example, the telecom operators can leverage the network assets to aggregate and resell the services of third-party clouds.

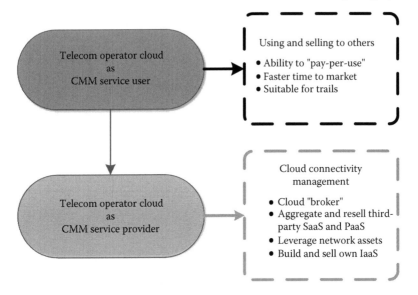

Figure 1.5 Telecom operator cloud. (From Yin, Z. et al., *IEEE Trans. Wireless Commun.*, 14, 4020, 2015. With permission.)

As shown in Figure 1.5, TOC is in an unique position of being as a cloud "broker" between the wireless networks and the third-party SPs, and can manage connectivity and offer flexibility in acquiring network resources on-demand and in real-time.

There are three major roles, namely, cloud connectivity, delivery of cloud-based capabilities, and leveraging network assets to enhance cloud offerings. This TOC model can align itself in the cloud value chain [14].

Pricing is an important issue in cloud computing. There are several studies about pricing schemes and algorithms for cloud services [2,9,19,21,31,37,39]. We can group them mainly into two categories, namely static pricing and dynamic pricing. In static pricing schemes, prices cannot be changed in a relatively short period, and the telecom operator cloud does not adapt to real-time congestion conditions, and there is no usage incentives [2,21,31,39]. By contrast, dynamic pricing can adjust the prices in nearly real-time in response to the observed network conditions [19,37]. In the mobile cloud environment that is considered in this research, end-users have the abilities to communicate with TOC in real-time. In addition, end-users can easily control their own usage on the individual devices and applications. Therefore, in this chapter, we adopt a dynamic pricing scheme.

1.3.1.2 Heterogeneous Wireless Networks with Small Cells

A promising approach to improve network performance in terms of capacity and energy efficiency is to use a multi-tier or hierarchical structure with small cells [22]. This architecture represents a novel wireless networking paradigm based on the idea of deploying short-range, low-power, and low-cost base stations, which operate in conjunction with macrocells.

Telecom operators have to deploy multiple wireless access networks with different technologies nowadays to meet the growing demands of users in regards to bandwidth and mobility [1]. HWNs is one of the solutions to make the handover between these technologies more transparent for the end-users, and to facilitate a more seamless experience for roaming. One of the key features in HWNs is to always provide the best data service and network connectivity to the end-users via different available wireless access networks when subjected to different interworking scenarios appearing throughout the time of handover and roaming procedures.

Figure 1.4 shows HWNs model, in which there is one *macrocell base station* (MBS) and multiple *small cell base stations* (SBSs). Each SBS is connected to the MBS via a broadband connection, such as a cable modem or *digital subscriber line* (DSL). The MBS and the small cells have a cognitive capability and can sense the channel state information. There are multiple *macrocell users* (MUs) and *orthogonal frequency-division multiple access* (OFDMA) technology is used. To simplify the analysis of this problem, without loss of generality, we assume that each small cell serves only one *small cell secondary user* (SSU). All the small cells are deployed, sparsely to avoid mutual interference. The macrocell and the small cells are considered to be perfectly synchronized, and the whole system is operated in a time-slotted manner.

As we consider that the macrocell and the small cell share the spectrum in the mobile network, there will be cross-tier interference between them, which will significantly affect the performance. To guarantee the QoS of end-users and reduce this effect, we introduce an interference price charged by MBS to protect itself from the SSU.

According to this price and the channel condition, the small cell will change the sub-bands they access and their transmission power.

1.3.1.3   A Game-Theoretic Approach

In this chapter, the problem of joint operations of telecom operator cloud and HWN is formulated as a three-stage Stackelberg game, which is shown in Figure 1.6. The main notations used in this chapter are listed in Table 1.1. Each third-party CMM SP is a leader that provides a cloud media service price x_r to the macrocell and small cells. All the MUs and the SSUs, which are playing the part of followers, decide the amount of media service from the CMM SPs to purchase according to the service price x_r in Stage I of the game. We measure the media service in *bits per second* (bps) to meet the end-user's media demand by guaranteeing performance. In Stage II, first, the MBS decides as a follower from which CMM SP to buy the media service, then it acts as a leader to offer an interference price y to the small cells to reduce the interference effect. In the last stage, each SBS

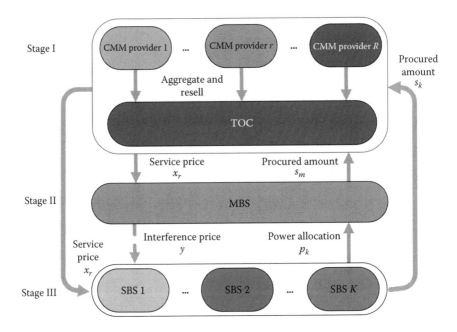

Figure 1.6   Three-stage Stackelberg game model. (From Yin, Z. et al., *IEEE Trans. Wireless Commun.*, 14, 4020, 2015. With permission.)

Table 1.1 Notations

Notation	Description
x_r	Cloud media service price
c_r	Cost of CMM SP r
y	Interference price
s_m	Amount of CMM service (MBS purchased from CMM SP r)
s_k	Amount of CMM service (SBS purchased from CMM SP r)
B_{rm}	Whether MBS m purchases service from provider r
B_{rk}	Whether SBS k purchases service from provider r
W	Transmission bandwidth of each channel
h_m	Channel gain between MBS and MU m
h_k	Channel gain between SBS k and SSU k
p_m	Transmit power of MBS to MU m
p_k	Transmit power of SBS k
σ_m	AWGN with zero mean and unit variation in Stage II
σ_k	AWGN with zero mean and unit variation in Stage III
g_{km}	Channel gain between the small cell k and the MU m
α	Parameter of tradeoff in Stage II
β	Parameter of tradeoff in Stage II
μ_k	Parameter of tradeoff in Stage III
λ_k	Parameter of tradeoff in Stage III

Source: Yin, Z. et al., *IEEE Trans. Wireless Commun.*, 14, 4020, 2015. With permission.

decides which CMM SP to buy the service from based upon the service price x_r and the interference price y charged by MBS.

Cloud Level Game

For the CMM SPs, we assume that each of them is selfish and independent of gaining the revenue as much as possible. Each CMM SP's profit depends on its own resource cost and the service price, as well as the price offered by the other SPs. For an arbitrary provider, we can formulate the utility function $U_r(x)$ as follows:

$$U_r(x) = (x_r - c_r)\left(s_m B_{rm} + \sum_{k=1}^{K} s_k B_{rk} \right). \tag{1.1}$$

The price vector $\mathbf{x}$ ($x = \{x_1, \ldots, x_r, \ldots, x_R\}$) denotes the prices offered by the SPs, and c_r denotes the cost of the CMM SP (e.g., the server cost, infrastructure cost, power usage, and the networking cost) [18]. The set R ($R = \{1, \ldots, r, \ldots, R\}$) denotes the number of the game players—CMM SPs. We assume that the cost of each CMM SP is different from the others. s_m and s_k denote MBS m and SBS k purchase the amount of CMM service from the CMM SP r. $B_{rm} \in \{0,1\}$ and $B_{rk} \in \{0,1\}$ denote whether MBS m and SBS k purchase the service from the SP r or not, where 1 means yes and 0 means no. The SP needs to find an optimal price to the MBS and SBSs in order to maximize its own revenue, which can be solved by the following problem:

$$\max_{x_r \geq c_r} U_r(x). \tag{1.2}$$

MBS Level Game and SBS Level Game

We need to consider both the transmission data rate and the computing resource consumption. To these ends, the MBS needs to limit the interference from small cells by offering the interference price y. We have MBS's net utility function defined as follows:

$$U_m(s_m, p_m, y) = \min\left(W\log_2\left(1 + \frac{p_m h_m^2}{\sigma_m^2 + \sum_{k=1}^{K} g_{km}^2 p_k}\right), s_m\right)$$

$$- \alpha x_r s_m B_{rm} + \beta y \sum_{k=1}^{K} g_{km}^2 p_k. \tag{1.3}$$

where

W denotes the transmission bandwidth of each channel

h_m denotes the channel gain from MBS m to its scheduled macrocell user, including the path-loss and the small-scale fading process

p_m denotes the transmit power of MBS m to its scheduled MU, here we assume this power is fixed

σ_m is AWGN with zero mean and unit variation

g_{km} denotes the channel gain between the small cell k and the scheduled MU served by MBS m including the path-loss and the small-scale fading process

y denotes the interference price

The set $K = \{1,\ldots,k,\ldots,K\}$ and $M = \{1,\ldots,m,\ldots,M\}$ denote the number of the SBSs and MBS in HWNs.

The min operator means that we should consider the smaller value of either demand or supply. We assume that we do not know the size of the capacity of MBS m and the amount of CMM service requested from the end-user scheduled by MBS m. If the size of the requirements is greater than the capacity of the MBS m, we have to choose the capacity to adapt the real environment, and vice versa.

α and β denote weights, which represent the tradeoff among the transmission rate, service cost, and interference revenue. Moreover, to make all the sub-formulas keep the operator symbols unchanged, we assume that α and β are greater than 0. The optimization problem for the MBS can be formulated as,

$$\max_{s_m>0,\ y\geq 0} U_m(s_m,y). \tag{1.4}$$

The net utility function for an arbitrary SBS k ($k \in \{1\ldots K\}$) can be defined as follows:

$$U_k(s_k,p_k) = \min\left(W\log_2\left(1+\frac{p_k h_k^2}{\sigma_k^2}\right),s_k\right) - \mu_k(x_r s_k B_{rk}) - \lambda_k y g_{km}^2 pk. \tag{1.5}$$

The symbol h_k denotes the channel gain between SBS k and SSU k, including the path-loss and the small-scale fading process, p_k denotes the transmit power of SBS k, and σ_k is AWGN. μ_k and λ_k denote weights, which represent the tradeoff among the transmission rate, service cost, and interference cost. Moreover, μ_k and λ_k are greater than 0.

The min operator means that we should consider the smaller value of either demand or supply. We assume that we do not know the size of the capacity of SBS k and the amount of CMM service requested from the end-user scheduled by SBS k. If the size of the requirements is greater than the capacity of the SBS k, we have to choose the capacity to adapt the real environment, and vice versa.

The optimization problem for an arbitrary small cell k can be formulated as,

$$\max_{s_k>0,\ p_k\geq 0} U_k(s_k,p_k). \tag{1.6}$$

1.3.2 Three-Stage Game Analysis

In this section, we analyze the proposed three-stage Stackelberg game. Then, we obtain the equilibrium to this game. Based on the description of the system, we know that each strategy will affect the others. Hence, we will use a backward induction method to solve this game. And an equilibrium point can always be obtained from the argument: starting from the final nodes, each player chooses a best reply given the (already determined) choices of all the players that move after him. This results in an equilibrium point also in each sub-game, whether that sub-game is reached or not. Such a point is called sub-game-perfect equilibrium, or backward induction equilibrium.

1.3.2.1 SBS Level Game Analysis

For maximizing the utility function of SBSs, each SBS will choose a proper CMM SP to purchase the CMM service according to the service price x_r, and the interference price y charged by MBS. For an arbitrary SBS k, its utility function will be defined in two cases: (1) $W \log_2 \left(1 + \left(p_k h_k^2 / \sigma_k^2\right)\right) \leq s_k$ and (2) $W \log_2 \left(1 + \left(p_k h_k^2 / \sigma_k^2\right)\right) > s_k$. The first case is the most common condition in real mobile cloud computing environment. In the most cases, the transmission rate of the CMM SP is larger than the transmission rate in the sub-band.

For the first case, we can obtain the utility function as follows:

$$U_k(s_k, p_k) = W \log_2 \left(1 + \frac{p_k h_k^2}{\sigma_k^2}\right) - \mu_k x_r s_k B_{rk} - \lambda_k y g_{km}^2 p_k. \tag{1.7}$$

When $1 - \mu_k x_r B_{rk} > 0$, in order to maximize U_k, we can use the decomposition theory. We decompose the optimization problem into two sub optimization problems, fix s_k then make p_k^* to be optimum to obtain the optimal s_k^*. So, its utility function is a concave function of p_k based on the definition in Equation 1.7, since

$$\frac{\partial^2 U_k}{\partial p_k^2} = -\frac{W h_k^4 (1 - \mu_k x_r B_{rk})}{\ln 2 (p_k h_k^2 + \sigma_k^2)} < 0. \tag{1.8}$$

Therefore, we can obtain the optimal power allocation strategy p_k^* as follows:

$$p_k^* = \left[\frac{W(1 - \mu_k x_r B_{rk})}{\ln 2 \lambda_k y g_{km}^2} - \frac{\sigma_k^2}{h_k^2}\right]^+. \tag{1.9}$$

Because the function of s_k is monotonic decreasing, then we know that it will achieve the maximum when s_k takes the minimum, that is,

$$
\begin{aligned}
s_k^* &= W \log_2\left(1 + \frac{p_k^* h_k^2}{\sigma_k^2}\right) \\
&= \left[W \log_2\left(\frac{h_k^2 W(1 - \mu_k x_r B_{rk})}{\ln 2 \lambda_k y g_{km}^2 \sigma_k^2}\right)\right]^+.
\end{aligned}
\tag{1.10}
$$

When $1 - \mu_k x_r B_{rk} < 0$, the utility function of p_k is a convex function, then we can obtain the minimum of the function as follows:

$$
p_k = \left[\frac{W(1 - \mu_k x_r B_{rk})}{\ln 2 \lambda_k y g_{km}^2} - \frac{\sigma_k^2}{h_k^2}\right]^+.
\tag{1.11}
$$

Hence, s_k can achieve optimal when $p_k = 0$. The utility function can be a monotonic decreasing function with s_k: $U_k(s_k, p_k) = -\mu_k x_r s_k B_{rk}$. So when $s_k^* = 0$, the function will get the maximum value, but this situation makes no sense in the real environment. Since an SSU applies for service from a mobile provider, its transmission rate should retain above zero.

For the second case, we can obtain the utility function of s_k as,

$$
\begin{aligned}
U_k(s_k) &= s_k - \mu_k(x_r s_k B_{rk}) - \lambda_k y g_{km}^2 p_k \\
&= (1 - \mu_k x_r B_{rk})s_k - \lambda_k y g_{km}^2 p_k.
\end{aligned}
\tag{1.12}
$$

When $1 - \mu_k x_r B_{rk} > 0$, $U_k(s_k)$ is a monotonic increasing function. To obtain the maximum value from this function, we know that $s_k^* = [W \log_2(1 + p_k^* h_k^2 / \sigma_k^2)]^+$. In here, we will keep the transmission power p_k unchanged to make s_k achieve the optimal value.

When $1 - \mu_k x_r B_{rk} < 0$, $U_k(s_k)$ is a monotonic decreasing function. So when $s_k^* = 0$, the function will get the maximum value, but this situation makes no sense in the real environment as well due to the reasons described earlier.

1.3.2.2 MBS Level Game Analysis

The MBS, in order to maximize its utility function, will firstly as a follower choose a proper CMM SP to purchase the CMM service based on the service price. Then it is as a leader to offer an interference price to the SBSs. We will obtain U_m, a function of transmission rate s_m and interference price y. In here, we consider the most common situation and also use the decomposition method to solve this problem. First, we keep the interference price y unchanged to get the optimal s_m^* to maximize U_m, then we obtain the desirable value of y. We assume that,

$$I_m = \sigma_m^2 + \sum_{k=1}^{K} g_{km}^2 p_k^*$$

$$= \sigma_m^2 + \sum_{k=1}^{K} g_{km}^2 \left[\frac{W(1 - \mu_k x_r B_{rk})}{\ln 2\lambda_k y g_{km}^2} - \frac{\sigma_k^2}{h_k^2} \right]^+. \tag{1.13}$$

In the MBS level, it will choose the CMM SP with the lowest service price x_r^*. For MBS m, its utility function will be defined in two cases:

1. $W \log_2 \left(1 + \frac{p_m h_m^2}{\sigma_m^2 + \sum_{k=1}^{K} g_{km}^2 p_k} \right) \leq s_m$

2. $W \log_2 \left(1 + \frac{p_m h_m^2}{\sigma_m^2 + \sum_{k=1}^{K} g_{km}^2 p_k} \right) > s_m$

For the first case, this situation is the most common condition in real mobile cloud computing environment. In most scenarios, the transmission rate of the CMM SP is larger than the sub-band transmission rate.

For the second case, it is a rare scene, but we cannot neglect it.

We obtain the utility function in the first case,

$$U_m(s_m, y) = W \log_2 \left(1 + \frac{p_m h_m^2}{I_m} \right) - \alpha x_r^* s_m + \beta y \sum_{k=1}^{K} g_{km}^2 p_k^*. \tag{1.14}$$

Because $\alpha, x_r \geq 0$, the utility function is monotonic decreasing of s_m, when s_m chooses the minimum value, the function will achieve the maximum value. Then, we can obtain the optimum of

$$s_m^* = \left[W \log_2 \left(1 + \frac{p_m h_m^2}{\sigma_m^2 + \sum_{k=1}^{K} g_{km}^2 \left(\frac{W(1-\mu_k x_r B_{rk})}{\ln 2\lambda_k y g_{km}^2} - \frac{\sigma_k^2}{h_k^2} \right)} \right) \right]^+. \tag{1.15}$$

And the utility function in the second condition is shown as follows:

$$U_m(s_m, y) = s_m(1 - \alpha x_r^*) + \beta y \sum_{k=1}^{K} g_{km}^2 p_k. \tag{1.16}$$

When $1 - \alpha x_r^* > 0$, to maximize the function, the value of s_m will be

$$s_m^* = \left[W \log_2 \left(1 + \frac{p_m h_m^2}{\sigma_m^2 + \sum_{k=1}^{K} g_{km}^2 p_k^*} \right) \right]^+. \tag{1.17}$$

When $1 - \alpha x_r^* \leqslant 0$, it is a monotonic decreasing function. To obtain the maximum of the function, $s_m = 0$, but this situation makes no sense in the real environment. Since an MU applies for service from a mobile provider, its transmission rate should retain above zero. And we have discussed that only in the condition of $1 - \mu_k x_r B_{rk} > 0$, parameter p_k^* can get the optimal value. Hence, we will continue the steps in the earlier default condition.

Due to the piece nature of interference price y, we present an indicator function,

$$D_k = \begin{cases} 1, & y < \frac{h_k^2 W (1 - \mu_k x_r B_{rk})}{\ln 2 \lambda_k \sigma_k^2 g_{km}^2}, \ \forall k \\ 0, & \text{otherwise.} \end{cases} \tag{1.18}$$

We can rewrite U_m that is shown as Equation 1.19.

$$U_m(y) = (1 - \alpha x_r^*) W \log_2 \left(1 + \frac{p_m h_m^2}{\sigma_m^2 + \sum_{k=1}^{K} g_{km}^2 D_k \left(\frac{W(1 - \mu_k x_r B_{rk})}{\ln 2 \lambda_k y g_{km}^2} - \frac{\sigma_k^2}{h_k^2} \right)} \right)$$
$$+ \beta y \sum_{k=1}^{K} g_{km}^2 D_k \left(\frac{W(1 - \mu_k x_r B_{rk})}{\ln 2 \lambda_k y g_{km}^2} - \frac{\sigma_k^2}{h_k^2} \right). \tag{1.19}$$

Because U_m is a piecewise function of y, we cannot solve it by derivation directly. When the value of D_k is given, we can obtain the function U_m as a continuous differentiable function. We let N_k as,

$$N_k = \frac{h_k^2 W (1 - \mu_k x_r B_{rk})}{\ln 2 \lambda_k \sigma_k^2 g_{km}^2}, \quad \forall k. \tag{1.20}$$

After having sorted all N_k in ascending order, like $N_1 \leqslant N_2 \leqslant \cdots \leqslant N_k$. Hence, we get K intervals $(0, N_1), (N_2, N_3), \ldots, (N_{k-1}, N_k)$. And by piecewise differentiating of function U_m in each interval, we can easily obtain it as concave except most N nondifferentiable points by analogizing as follows:

$$\frac{\partial U_m(y)}{\partial y} = (1 - \alpha x_r^*) \sum_{k=1}^{K} \frac{W^2 p_m h_m^2 (1 - \mu_k x_r B_{rk})}{(\ln 2)^2 y^2 I_m(y)(p_m h_m^2 + I_m(y))\lambda_k}$$

$$- \beta \sum_{k=1}^{K} g_{km}^2 \frac{\sigma_k^2}{h_k^2}. \tag{1.21}$$

Based on Equation 1.21, we know $\frac{\partial^2 U_m(y)}{\partial y^2}$ is shown as Equation 1.22.

$$\frac{\partial^2 U_m(y)}{\partial y^2} = -(1 - \alpha x_r^*) \sum_{k=1}^{K} \frac{W^2 p_m h_m^2 (1 - \mu_k x_r B_{rk})}{(\ln 2)^2 \lambda_k y^3 (p_m h_m^2 I_m + I_m^2)^2}$$

$$\times \left(p_m h_m^2 I_m + (p_m h_m^2 + 2I_m) \left(\sigma_m^2 - \sum_{k=1}^{K} g_{km}^2 \frac{\sigma_k^2}{h_k^2} \right) \right) < 0. \tag{1.22}$$

Therefore, $U_m(y)$ is a concave function. And this noncooperative interference price game is a concave game. According to the formulation Equation 1.19, this game exists at least one Nash equilibrium.

Then, from Equation 1.22, we can know that $\frac{\partial U_m(y)}{\partial y}$ is a strictly monotonic decreasing function of y.

Obviously, we find that,

$$\lim_{y \to \infty} I_m = \sigma_m^2 - \sum_{k=1}^{K} g_{km}^2 \frac{\sigma_k^2}{h_k^2} > 0, \tag{1.23}$$

then,

$$\lim_{y \to \infty} \frac{\partial U_m(y)}{\partial y} = -\beta \sum_{k=1}^{K} g_{km}^2 \frac{\sigma_k^2}{h_k^2} < 0. \tag{1.24}$$

For the case of $y \to 0$ we can obtain that,

$$\lim_{y \to 0} \frac{\partial U_m(y)}{\partial y} = \infty > 0. \tag{1.25}$$

Therefore, the utility function U_m is first increasing with the interference price y, then at the certain point begins to decrease with y. So this utility function is a concave function without some nondifferentiable points $N_k, k \in (1,\ldots,K)$. That is, this noncooperative competitive game exists at least one Nash equilibrium. We can find this optimal value of y in each interval by multiple methods (e.g., a binary search algorithm and a gradient-based algorithm).

1.3.2.3 CMM SP Level Game Analysis

Bertrand game is a popular tool to model competition among firms (sellers) that set prices and their customers (buyers) that choose quantities at the price set. Bertrand game has been successfully applied in cognitive radio networks, and other areas [42]. In our scheme, we use Bertrand game to model the competition among CMM SPs. We assume that each CMM SP is independent, acts selfishly, and the target is to gain as much revenue as possible. If the CMM SPs act noncooperatively, it will lead a monopoly situation. All the CMM SPs are eager to set their service prices as the same, and try to maximize their whole profits. The profit of an arbitrary CMM SP r depends not only on the service price x_r and the cost c_r, but also on the service prices x_{-r} offered by the other cloud SPs [30].

Each CMM SP decides its action independently and simultaneously. And the CMM SP with the lowest price will occupy the entire service market. Hence, every CMM SP tries to reduce its service price until hitting the bottom with zero profit. As discussed in Section 1.3.1, the set of the game players is $\mathbf{R} = \{1,\ldots,r,\ldots,\mathcal{R}\}$, the strategy set is x_r, and the payoff function of the CMM SP is U_r. The NE of this problem gives the set of prices such that neither CMM SP can increase its net profit U_r by unilaterally changing the price. Without loss of generality, let the cost set in an ascending order $c_1 < c_2 < \ldots < c_{\mathcal{R}}$.

Proposition 1.1 *The NE of the proposed homogeneous Bertrand game with multiple CMM SPs is shown as follows:*

$$x^* = \{x_1^*, c_2, c_3, \ldots, c_{\mathcal{R}}\}, \tag{1.26}$$

where x_1^* denotes the price strategy of the first CMM SP at the Nash-equilibrium, which can be formulated as,

$$x_1^* = \arg \max_{c_1 \le x_1 < c_2} U_1^M(x_1), \tag{1.27}$$

where U_1^M is the utility function of the provider 1 when it supplies the whole market, shown as Equation 1.28.

$$U_1^M = (x_1 - c_1)\left(B_{rm}\left[W \log_2\left(1 + \frac{p_m h_m^2}{I_m}\right)\right]^+ \right.$$
$$\left. + \sum_{k=1}^{K} B_{rk}\left[W \log_2\left(\frac{h_k^2 W(1 - \mu_k x_1 B_{rk})}{\ln 2\lambda_k y g_{km}^2 \sigma_k^2}\right)\right]^+\right). \tag{1.28}$$

Proof: By observing Equation 1.19, we know that U_1^M is the utility function of x_1. Assume in a Bertrand condition, there are only two CMM SPs, the provider 1 and the provider 2 in the competition. The costs are defined as c_1 and c_2, and we let $c_1 < c_2$. According to the assumption about the Bertrand competition model, both providers have the incentives to reduce their service prices down to their own margin cost to undercut the other and capture the whole market to almost double its profit. That is to say, if one provider sets its price equals to the marginal cost, and the other provider tries to raise its service price over the cost, then it will earn nothing. Since all the end-users would purchase the service from the provider that is still setting the competitive price. If one provider has a minor average cost, it will charge the highest price that is lower than the average cost of the other one and takes all the business. So provider 1 has the incentive to make its price between the set of $[c_1, c_2)$ to maximize its own profit.

Though it is irrational to set the price below the marginal cost, the two CMM SPs would make the prices lower than their own monopoly price. If the provider 2 presents a high enough price, the provider 1 can definitely ignore the affection of provider 2 and set its price by the optimal monopoly price.

Due to the piecewise property about x_1, to maximize the utility function U_1^M, for all $k \in \{1, 2, \ldots, K\}$, we introduce the indicator function as follows:

$$V_m = \begin{cases} 1, & x_1 < \dfrac{W - \ln 2\lambda_k y\left[\frac{1}{K}(p_m h_m^2 - \sigma_m^2) + g_{km}^2 \frac{\sigma_k^2}{h_k^2}\right]}{W \mu_k B_{rk}}, \forall k \\ 0, & \text{otherwise.} \end{cases} \tag{1.29}$$

$$V_k = \begin{cases} 1, & x_1 < \frac{Wh_k^2 - \ln 2\lambda_k y g_{km}^2 \sigma_k^2}{Wh_k^2 \mu_k B_{rk}}, \forall k \\ 0, & \text{otherwise.} \end{cases} \tag{1.30}$$

Thus by substituting Equations 1.29 and 1.30 into Equation 1.28, we can know that the utility function is a piecewise function about x_1 due to those indicator functions V_m and V_k, which cannot obtain the derivative directly to solve x_1. However, if determined V_m and V_k, we still can get a continue differentiable function.

Therefore we let,

$$F_m = \frac{W - \ln 2\lambda_k y \left[\frac{1}{K}(p_m h_m^2 - \sigma_m^2) + g_{km}^2 \frac{\sigma_k^2}{h_k^2} \right]}{W \mu_k B_{rk}}, \tag{1.31}$$

$$F_k = \frac{Wh_k^2 - \ln 2\lambda_k y g_{km}^2 \sigma_k^2}{Wh_k^2 \mu_k B_{rk}}. \tag{1.32}$$

Then sort all $F_1, F_2, \ldots, F_m$, and F_k in an ascending order $F_1 \leq F_2 \leq \cdots F_m \leq F_k$, without loss of generality. Hence we can get K intervals, $[0, F_1), (F_2, F_3), \ldots, (F_m, F_k)$. By piecewise differentiating of the utility function in the first interval, we assume that $0 < x_1 < F_1$, the second derivative of $U_1^M(x_1)$ is shown as Equation 1.33.

$$\frac{\partial^2 U_1^M(x_1)}{\partial x_1^2} = 2 \left\{ \sum_{k=1}^{K} \frac{W^2 \mu_k B_{rk} B_{rm} p_m h_m^2}{(\ln 2)^2 \lambda_k y I_m (p_m h_m^2 + I_m)} - \sum_{k=1}^{K} \frac{B_{rk}^2 W \mu_k}{\ln 2(1 - \mu_k x_1 B_{rk})} \right\}$$
$$+ (x_1 - c_1) \left\{ \sum_{k=1}^{K} \frac{W^3 \mu_k^2 B_{rk}^2 B_{rm} p_m h_m^2 (p_m h_m^2 + 2I_m)}{(\ln 2)^3 \lambda_k y I_m^2 (p_m h_m^2 + I_m)^2} \right.$$
$$\left. - \sum_{k=1}^{K} \frac{W B_{rk}^3 \mu_k^2}{\ln 2(1 - \mu_k x_1 B_{rk})^2} \right\}. \tag{1.33}$$

We have known the definition of I_m from Equation 1.13, for substituting it into Equation 1.33; we can obtain that the second derivative $\frac{\partial^2 U_1^M(x_1)}{\partial x_1^2}$ is less than 0, which means that $\frac{\partial U_1^M(x_1)}{\partial x_1}$ is a monotonic decreasing function of x_1. Obviously, we obtain the function given in Equation 1.34.

$$\frac{\partial U_1^M(x_1)}{\partial x_1} = B_{rm}W\log_2\left(1 + \frac{p_m h_m^2}{I_m}\right) + \sum_{k=1}^{K} B_{rk}W\log_2\frac{Wh_k^2(1 - \mu_k x_1 B_{rk})}{\ln 2\lambda_k y g_{km}^2 \sigma_k^2}$$

$$+ (x_1 - c_1)\left\{\sum_{k=1}^{K} \frac{W^2\mu_k B_{rk}B_{rm}p_m h_m^2}{(\ln 2)^2\lambda_k y I_m(p_m h_m^2 + I_m)}\right.$$

$$\left. - \sum_{k=1}^{K} \frac{B_{rk}^2 W\mu_k}{\ln 2(1 - \mu_k x_1 B_{rk})}\right\}. \tag{1.34}$$

From Equation 1.34, we have,

$$\lim_{x_1 \to 0} \frac{\partial U_1^M(x_1)}{\partial x_1} > 0. \tag{1.35}$$

When $x \to F_1$, we can get two cases as follows:

$$(1)\ \lim_{x \to F_1} \frac{\partial U_1^M(x_1)}{\partial x_1} \geq 0;$$

$$(2)\ \lim_{x \to F_1} \frac{\partial U_1^M(x_1)}{\partial x_1} < 0. \tag{1.36}$$

For the first case, the utility U_1^M is strictly monotonic increasing about x_1 at the initial interval $[0, F_1)$. Therefore, for the second case, we know that the utility function U_1^M first climbs up with x_1. After reaching the optimal point, it drops down with x_1. Hence, the utility function U_1^M is a concave function at the first interval, and it is easily to prove the utility function U_1^M is the concave function at the other intervals. That is to say, for $x_1 < F_k$, the utility function is a concave function without the most K non-differentiable points at $F_1, F_2, \ldots, F_m$, and F_k. We can solve x_1 at each interval by many methods (e.g., a binary search algorithm and a gradient based algorithm).

1.3.2.4   Service Allocation Iteration Algorithm

It is important to investigate the uniqueness and the existence of the Stackelberg equilibrium. In the duopoly case, the convexity of the follower's reaction function is essential for uniqueness of the Stackelberg equilibrium [40]. Hence, we will prove that for our model of Stackelberg game has an unique equilibrium.

Theorem 1.1 *The unique Nash equilibrium exists in the proposed Stackelberg game.*

Proof: In our Stackelberg game model, each stage has its flawless equilibrium in a Nash equilibrium, respectively: the service price strategies x^* in Equation 1.26 offered by the CMM SPs, the service allocation strategy s_m^*, the interference price, and the service allocation strategy s_k^* in Equation 1.10. Because we have proven that each stage exists a perfect equilibrium in a Nash equilibrium, the Nash equilibrium of the proposed Stackelberg game model exists. We also know that the subgame perfect equilibrium in each stage is unique. Therefore, the total Stackelberg Nash equilibrium is unique.

To get the Nash equilibrium of the three-stage Stackelberg game, we use a backward induction to solve the problem and present the service iteration algorithm.

In the aforementioned method, we defined the other CMM SPs' strategies as $x_{-r} = (x_1, x_2, \ldots, x_{r-1}, x_{r+1}, \ldots, x_R)$. When x_{-r} is given at iteration $t - 1$, we present the best response function $\mathcal{B}_r(x_{-r}[t-1])$ of CMM SP x_r at the iteration t to maximize its total revenue. The condition $\|x[t] - x[t-1]\| / \|x[t-1]\| \leq \varepsilon$ is the stop criteria. In the proposed algorithm, the MBS decides the interference price y offering to SBSs and the amount of service purchased from the CMM SPs based on the service price x_r. The SBS then allocates the power. The algorithm will stop until the service price x_r converges.

In practice, the proposed iterative algorithm to obtain the three-stage Stackelberg game equilibrium can be implemented as follows:

1. The CMM SPs randomly offer the CMM service price to the MBS and SBSs.

2. The MBS receives the channel state information from the SBSs and the MUs.

3. The MBS decides which provider to purchase the service from and the interference price offering to the SBSs.

4. The SBSs perform their power allocation.

5. The CMM SPs update their prices and repeat steps 2, 3, and 4 until the prices converge.

To ensure the convergence to the NE for the Algorithm 1.1, some sufficient conditions have been proposed in the existing literature.

Algorithm 1.1 Service Allocation Iteration Algorithm

Initialization

Initialize the cloud computing service prices, that is, for each CMM SP r, randomly offers the service price x_r, where $x_r \geq 0$.

Repeat Iterations

(a) The MBS offers the interference price y to the SBSs and decides which CMM SP to purchase the service from based upon x_r and the amount of computing service.

(b) Each SBS performs its service allocation.

(c) CMM SPs update their prices: $x_r[t] = B_r(x_{-r}[t-1])$.

(d) *Until:* $\|x[t] - x[t-1]\|/\|x[t-1]\| \leq \varepsilon$.

End Iteration

The convergence condition was first provided by [46] for the two-user case and extended for the N-users in [11]. Moreover, the conditions of the convergence were further proved in [24,36]. However, as the pricing factor x_r is recalculated in every iteration, the algorithm is actually a time-varying over iterations. Thus, the fixed-point theorem proposed in [24,36] may not be applied here. The convergence proof under a time-varying mapping function is a challenging problem and will be left for the future work. Nevertheless, convergence has always been observed in our simulations.

1.3.3 Simulation Results and Discussions

In this section, we use computer simulations to evaluate the performance of the proposed scheme by MATLAB® software. All the simulations are executed on a laptop featured with Windows 7, Intel Core i5 2.6 GHz CPU, 8GB memory, and MATLAB R2012a. The main system parameters in HWNs are adopted from 3GPP [1], as listed in Table 1.2. In the simulations, we assume that there are two CMM SPs. The MUs and small cells are located in the macrocell randomly, and small cells deploy sparsely with each other at 50–150 m far from the MBS. Following [1], we set the path loss between SSU and MBS as $15.3 + 37.6\log_{10}(D_{ms} + L_{ow})$, and the path loss between SSU and SBS as $46.86 + 20\log_{10}(D_{ss} + L_{iw})$. D_{ms} is the distance between the MBS and the SSU. D_{ss} is the distance between the SBS and the SSU. L_{ow} means the penetration loss of exterior wall, and L_{iw} means the penetration

Table 1.2 Simulation Parameters and Assumptions for Performance Evaluations

Parameters	Values/Assumptions
Deployment of stand-alone small cells	Randomly deployed at 50–150 m far from the MBS
Path loss: MBS ↔ SSU	$15.3 + 37.6 \log_{10} D_{ms} + L_{ow}$
Path loss: SSU ↔ SBS	$46.86 + 20 \log_{10} D_{ss} + L_{iw}$
Penetration loss of exterior wall L_{ow}	20 dB

Source: Yin, Z. et al., *IEEE Trans. Wireless Commun.*, 14, 4020, 2015. With permission.

Note: D_{ms} is the distance between the MBS and the SSU; D_{ss} is the distance between the SBS and the SSU.

loss of the interior wall. They are set as 20 and 10 dB, respectively. We can also find the parameters of the small scale and shadow fading in [1]. The others are shown as follows. The transmission bandwidth W is 5 MHz in the MBS and the transmission power is fixed as 46 dBm from [1]. The general parameters are set as, $\mu_k = 0.05$, $\lambda_k = 1$, $\alpha = 0.03$, $\beta = 10$, B_{rk} and $B_{rm} = 1$.

First, we evaluate the performance of the CMM service purchased by BSs with various lowest prices. Figure 1.7 shows that, with the increase of the lowest service price x_r, SBS_1 and SBS_2 have to decrease their transmission rate by performing the energy-efficiency power allocation. Due to the fixed transmission power of the MBS, it performs an increasing trend with x_r^*. The reason is that, by increasing x_r^*, the transmission power of SBS decreases as well, which in turn leads to the decrease of interference between the MBS and the SBS. Hence, the transmission rate of the MBS has the same variation trend with the service price x_r^*. When the service price is too high to afford for the SSUs, the SBSs will lower their transmission rate step by step until they stop transmitting anything. Because the SBSs stop their transmission, the SSUs choose another method to receive the CMM service due to the high service prices. The MBS will reach its highest level of the transmission rate shown in Figure 1.7. The shape of the curve can change with the parameters. However, the insight remains the same in figure.

Then, we compare the CMM service purchased by one SBS with three different values of x_r^*. Figure 1.8 shows that the SBS tries to

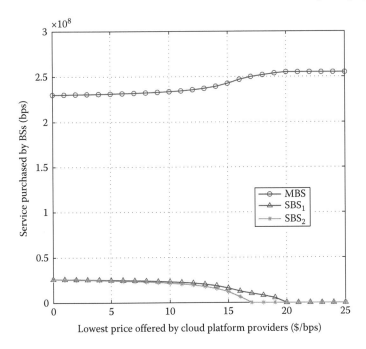

Figure 1.7 Service allocation with various lowest prices offered by the CMM service providers. (From Yin, Z. et al., *IEEE Trans. Wireless Commun.*, 14, 4020, 2015. With permission.)

reduce its interference cost y by decreasing its transmission rate s_k when given the value of service price offered by the CMM SP. Figure 1.8 also shows that, in the condition of the same interference price, the lower service price offered by the CMM SPs, the higher transmission rate we can obtain.

We also study the service allocation in the MBS with various service prices offered by the CMM SPs in Figure 1.9. The interference between MBS and SBSs will be reduced by the increasing trend of the service price x_r. That is because the transmission power p_m is fixed in the MBS, the transmission rate of MBS will increase with the service price until it reaches the highest level, then the interference turns to zero. We can find how the value of α effects the shape of the transmission rate s_k in MBS, the smaller the value of α, the higher level of transmission rate we can obtain. As α increases, there is tradeoff among the transmission rate, service cost, and interference revenue in the MBS.

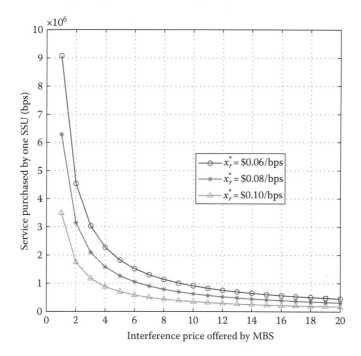

Figure 1.8   Service allocation in one SBS with various lowest service prices offered by the MBS. (From Yin, Z. et al., *IEEE Trans. Wireless Commun.*, 14, 4020, 2015. With permission.)

Figure 1.10 shows that the utility function of SBS is a concave function, which is proven in Section 1.3.2.1. In this figure, we know that the utility of SBS, first, increases as the power. After it reaches the optimal level, the utility of SBS begins to decrease, because the gain of the transmission rate cannot offset the increase trend of the service cost and interference price. This figure also tells us that the higher the interference price is, the lower utility of SBS will be.

In addition, we find that the transmission rate of the MBS correspondingly increases with the interference price in Figure 1.11. That is because the higher interference price forces the SBSs to reduce their transmission power. We also observe that the higher the service price x_r^* is, the lower the interference from the SBSs will be. Hence, the transmission rate will reach a higher level in the MBS with a more expensive service price offered by the CMM SPs.

Figure 1.12 shows the convergence of the proposed Stackelberg equilibrium iteration algorithm. We evaluate the performance of the

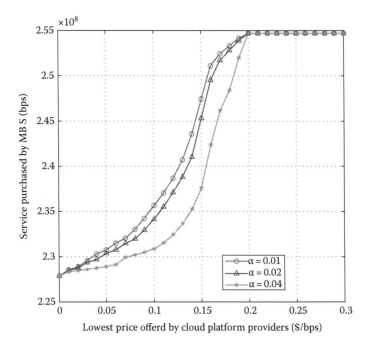

Figure 1.9 Service allocation in the MBS with various values of α parameter. (From Yin, Z. et al., *IEEE Trans. Wireless Commun.*, 14, 4020, 2015. With permission.)

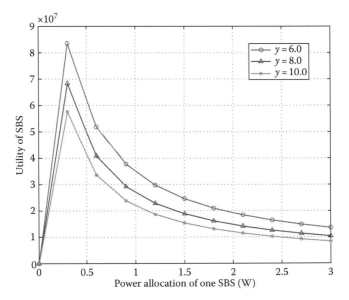

Figure 1.10 Utility function of one SBS. (From Yin, Z. et al., *IEEE Trans. Wireless Commun.*, 14, 4020, 2015. With permission.)

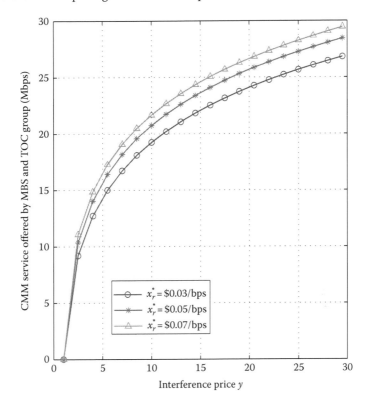

Figure 1.11   The performance of CMM service in the MBS. (From Yin, Z. et al., *IEEE Trans. Wireless Commun.*, 14, 4020, 2015. With permission.)

CMM service over the iteration steps. From the figure, we can observe the service of MBS and SBSs can converge after a few iteration steps because of the convergence of the service price x_r^*. Hence, we will obtain the NE by the algorithm.

Then, we study the frame rate (frames per second) and latency of CMM service in the mobile cloud computing environment under two different configurations: the local servers and the remote cloud computing servers over a 100 Mbps network with the output viewed through the *virtual network computing* (VNC) protocol, which have the different latency due to the different distances among the remote servers. Figure 1.13 demonstrates that a high frame rate provides the illusion of smoothness to an end-user. Even a modest latency of 33 ms can cause the frame rate to drop dramatically from that experienced with a local server. Although the VNC protocol strives to keep

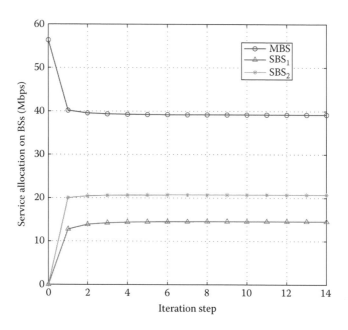

Figure 1.12   CMM service iteration step. (From Yin, Z. et al., *IEEE Trans. Wireless Commun.*, 14, 4020, 2015. With permission.)

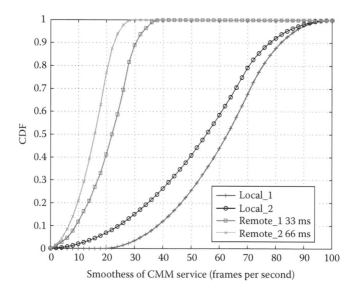

Figure 1.13   Network latency hurts interactive performance even with good bandwidth. (From Yin, Z. et al., *IEEE Trans. Wireless Commun.*, 14, 4020, 2015. With permission.)

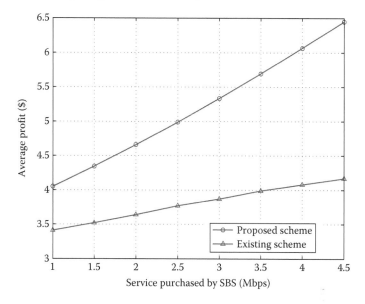

Figure 1.14 The average profit versus service allocation on one SBS. (From Yin, Z. *et al.*, *IEEE Trans. Wireless Commun.*, 14, 4020, 2015. With permission.)

the frame rate at an acceptable level, it offers sluggish interaction. Hence, the user experience is considerably poorer than that for the local media service interaction.

Finally, the average profit for our proposed scheme compared with an existing scheme [43] is evaluated in Figure 1.14. In the existing scheme, multiple services prices are offered to the SUs. This service pricing scheme adopts the nonincentive compatible differentiated type, which can show the theoretical upper bound of the overall profit of this method. The transmission rate is controlled between 1 and 4.5 Mbps. We recalculate the profits of the CMM SPs 1000 times at each 0.5 Mbps interval, then obtain the average profits. In the figure, we compare the average profit in our scheme and the existing scheme based on the CMM service purchased by the SBSs. We can find that the profits in the two schemes both have the growing trends with the increasing CMM service purchased by the SBS. Simulation results show that the proposed scheme can gain more profit than the existing one in the same service allocation condition. That is because the existing scheme does not consider the dynamic resource allocation method.

1.4   Conclusions

In this chapter, we have studied the issues that arise when jointly considering the operations of cloud and wireless networks in a mobile cloud computing environment with a telecom operator cloud. We introduced a system model, which jointly considers the CMM SPs and HWNs with small cells. Multiple CMM SPs offer CMM service prices to the heterogeneous networks. Then, the MBS and SBSs adjust the amount of service they procured by performing resource allocation. We formulated the problems of determining a CMM service price, wireless power allocation, and interference management as a three-level Stackelberg game. We also presented an interference price to measure and mitigate the cross-tier interference between the macrocell and small cells. The MBS is allowed to protect its own users by charging the SBSs. At the CMM SP's level, we proposed a homogeneous Bertrand game with asymmetric costs to model the CMM service decisions and used a backward induction method to solve the whole model. Finally, we presented an iteration algorithm to obtain the equilibrium of the Stackelberg Game. Simulation results have been presented to show that the dynamics of cloud operations have a significant impact on the HWN, and joint optimization is necessary for the operations of cloud and wireless networks. This is due to unique dynamics tied with cloud, CMM services and wireless networks. By jointly optimizing the operations of clouds and wireless networks, the proposed scheme can significantly improve the performance of the mobile cloud computing systems.

References

1. 3GPP. Further advancements for E-UTRA physical layer aspects (release 9). 3GPP TR36.814, 2010.

2. M. Andrews, U. Ozen, M. I. Reiman, and Q. Wang. Economic models of sponsored content in wireless networks with uncertain demand. In *Proceedings of IEEE INFOCOM'13*, Turin, Italy, pp. 3213–3218. IEEE, 2013.

3. H. Balakrishnan, V. N. Padmanabhan, S. Seshan, and R. H. Katz. A comparison of mechanisms for improving TCP performance over wireless links. *IEEE/ACM Trans. Netw.*, 5(6):756–769, Dec. 1997.

4. H. Ballani, T. Karagiannis, and P. Costa. Towards predictable datacenter networks. In *Proceedings of ACM SIGCOMM'2011*, Toronto, Ontario, Canada, pp. 242–253, 2011.

5. T. Benson, A. Akella, A. Shaikh, and S. Sahu. Cloudnaas: A cloud networking platform for enterprise applications. In *Proceedings of the ACM Symposium on Cloud Computing*, Cascais, Portugal, p. 8. ACM, 2011.

6. S. Bu, F. R. Yu, and Y. Cai. When the smart grid meets energy-efficient communications: Green wireless cellular networks powered by the smart grid. *IEEE Trans. Wireless Commun.*, 11(8):3014–3024, 2012.

7. B. Calder and J. Wang. Windows Azure storage: A highly available cloud storage service with strong consistency. In *Proceedings of 23rd ACM SOSP'2011*, pp. 143–157, New York. ACM, 2011.

8. R.-S. Chang, J. Gao, V. Gruhn, J. He, G. Roussos, and W.-T. Tsai. Mobile cloud computing research—Issues, challenges and needs. In *IEEE Seventh International Symposium on Service Oriented System Engineering (SOSE)*, San Francisco, CA, pp. 442–453, Mar. 2013.

9. Y.-F. Robin Chen and R. Jana. Speedgate: A smart data pricing testbed based on speed tiers. In *Proceedings of IEEE INFOCOM'13*, Turin, Italy, pp. 3195–3200, 2013.

10. Z. Chen and D. Wu. Rate-distortion optimized cross-layer rate control in wireless video communication. *IEEE Trans. Circuits System Video Technol.*, 22(3):352–365, Mar. 2012.

11. S. T. Chung, S. J. Kim, J. Lee, and J. M. Cioffi. A game-theoretic approach to power allocation in frequency-selective Gaussian interference channels. In *Proceedings of IEEE International Symposium on Information Theory*, Yokohama, Japan, pp. 316–316, 2003.

12. A. Corradi, M. Fanelli, and L. Foschini. VM consolidation: A real case based on openstack cloud. *Future Gener. Comp. Syst.*, 32:118–127, 2014.

13. P. Costa, M. Migliavacca, P. Pietzuch, and A. L. Wolf. NaaS: Network-as-a-service in the cloud. In *Proceedings of the Second USENIX Hot-ICE'12*, San Jose, CA, vol. 12, pp. 1–1, 2012.

14. ERICSSON. The telecom cloud opportunity. In ERICSSON Discussion Paper, http://www.ericsson.com/, accessed on Sept. 26, 2013., Mar. 2012.

15. N. Fernando, S. Loke, and W. Rahayu. Mobile cloud computing: A survey. *Elsevier Future Gener. Comp. Syst.*, 29(1):84–106, Jan. 2013.

16. N. Fernando, S. Loke, and W. Rahayu. Mobile cloud computing: A survey. *Future Gener. Comp. Syst.*, 29(1):84–106, 2013.

17. A. Ghosh et al. Heterogeneous cellular networks: From theory to practice. *IEEE Commun. Mag.*, 50(6):54–64, June 2012.

18. A. Greenberg, J. Hamilton, D. A. Maltz, and P. Patel. The cost of a cloud: Research problems in data center networks. *ACM SIG-COMM Comput. Commun. Rev.*, 39(1):68–73, 2008.

19. S. Ha, S. Sen, C. Joe-Wong, Y. Im, and M. Chiang. Tube: Time-dependent pricing for mobile data. In *Proceedings of the ACM SIG-COMM'12*, Helsinki, Finland, pp. 247–258. ACM, 2012.

20. S. Hart. Evolutionary dynamics and backward induction. *Games Econ. Behav.*, 41(2):227–264, 2002.

21. Y. Hayel and B. Tuffin. A mathematical analysis of the cumulus pricing scheme. *Comp. Netw.*, 47(6):907–921, 2005.

22. J. Hoadley and P. Maveddat. Enabling small cell deployment with HetNet. *IEEE Trans. Wireless Commun.*, 19(2):4–5, April 2012.

23. D. Huang et al. Mobile cloud computing. *IEEE MMTC E-Lett.*, 6(10):27–31, 2011.

24. J. Huang, R. Cendrillon, M. Chiang, and M. Moonen. Autonomous spectrum balancing (ASB) for frequency selective interference channels. In *Proceedings of IEEE International Symposium on Information Theory*, Seattle, WA, pp. 610–614, 2006.

25. J. Huang, Z. Li, M. Chiang, and A. K. Katsaggelos. Joint source adaptation and resource allocation for multi-user wireless video streaming. *IEEE Trans. CSVT*, 18(5):582–595, 2008.

26. Juniper Research. Mobile cloud: Smart device strategies for enterprise and consumer markets 2011-2016. Technical report, http://juniperresearch.com, July 2011 (accessed July 1, 2015).

27. R. Kokku, R. Mahindra, H. Zhang, and S. Rangarajan. NVS: A substrate for virtualizing wireless resources in cellular networks. *IEEE/ACM Trans. Netw.*, 20(5):1333–1346, 2012.

28. D. Kovachev, D. Renzel, R. Klamma, and Y. Cao. Mobile community cloud computing: Emerges and evolves. In *Proceedings of the IEEE 11th MDM'10*, Kansas City, MI, pp. 393–395, 2010.

29. K. Kumar and L. Yung-Hsiang. Cloud computing for mobile users: Can offloading computation save energy? *Computer*, 43(4):51–56, 2010.

30. A. Ledvina and R. Sircar. Oligopoly games under asymmetric costs and an application to energy production. *Math. Financial Econ.*, 6(4):261–293, 2012.

31. P. Marbach. Analysis of a static pricing scheme for priority services. *IEEE/ACM Trans. Netw.*, 12(2):312–325, 2004.

32. Nokia. Network virtualization: Enabling novel business models in a dynamic market. Technical report, 2012. www.nokia.com (accessed July 1, 2015).

33. ONF. ONF white paper: Software-defined networking: The new norm for networks. Technical report, ONF, Palo Alto, CA, 2012.

34. X. Renchao, F. R. Yu, J. Hong, and L. Yi. Energy-efficient resource allocation for heterogeneous cognitive radio networks with femtocells. *IEEE Trans. Wireless Commun.*, 11(11):3910–3920, 2012.

35. Z. Sanaei, S. Abolfazli, A. Gani, and M. Shiraz. Sami: Service-based arbitrated multi-tier infrastructure for mobile cloud computing. In *Proceedings of the ICCC'12*, Beijing, China, pp. 14–19. IEEE, 2012.

36. G. Scutari, D. P. Palomar, and S. Barbarossa. Asynchronous iterative waterfilling for Gaussian frequency-selective interference channels: A unified framework. In *Proceedings of Inform. Theory Appl. Workshop, 2007*, San Diego, CA, pp. 349–358, 2007.

37. S. Sen, C. Joe-Wong, S. Ha, and M. Chiang. A survey of smart data pricing: Past proposals, current plans, and future trends. *ACM Comput. Surv.*, 2013.

38. S. Sezer, S. Scott-Hayward, P. K. Chouhan, B. Fraser, D. Lake, J. Finnegan, N. Viljoen, M. Miller, and N. Rao. Are we ready for SDN? Implementation challenges for software-defined networks. *IEEE Commun. Mag.*, 51(7):36–43, July 2013.

39. S. Shakkottai and R. Srikant. Economics of network pricing with multiple ISPs. *IEEE/ACM Trans. Netw.*, 14(6):1233–1245, 2006.

40. H. D. Sherali, A. L. Soyster, and F. H. Murphy. Stackelberg-Nash-Cournot equilibria: Characterizations and computations. *Operations Res.*, 31(2):253–276, 1983.

41. B. Wang, Z. Han, and K. J. R. Liu. Distributed relay selection and power control for multiuser cooperative communication networks using Stackelberg game. *IEEE Trans. Mobile Comput.*, 8:975–990, Jul. 2009.

42. B. Wang, Y. Wu, and K. J. Liu. Game theory for cognitive radio networks: An overview. *Comput. Netw.*, 54(14):2537–2561, 2010.

43. C.-Y. Wang and H.-Y. Wei. Profit maximization in femtocell service with contract design. *IEEE Trans. Wireless Commun.*, 12(5):1978–1988, 2013.

44. S. Wang and S. Dey. Rendering adaptation to address communication and computation constraints in cloud mobile gaming. In *Proceedings of IEEE Globecom'2010*, Miami, FL, pp. 1–6, 2010.

45. X. Wang, P. Krishnamurthy, and D. Tipper. Wireless network virtualization. In *Proceedings of ICNC'13*, San Diego, CA, pp. 818–822. IEEE, 2013.

46. W. Yu. Competition and cooperation in multi-user communication environments. PhD thesis, Stanford University, Stanford CA, 2002.

47. Z. Yin, F. Richard Yu, S. Bu, and Z. Han. Joint cloud and wireless networks operations in mobile cloud computing environments with telecom operator cloud. *IEEE Wireless Commun.*, 14(7):4020–4033, 2015.

CHAPTER 2

Toward Energy-Efficient Task Execution in Mobile Cloud Computing

Yonggang Wen, Weiwen Zhang, and Kyle Guan

CONTENTS

2.1   Introduction

Mobile cloud computing has been hailed as an effective solution to extend capabilities of resource-poor mobile devices by application off-loading [12,14,47]. Application off-loading allows the mobile devices to run computation-intensive applications. Such applications include virus scanning [10], face recognition [11], image retrieval [29], optical character recognition [30], mobile gaming [45], rescue missions on robotics [18], and healthcare systems [48]. However, it is not always energy-efficient to off-load mobile applications to the cloud for execution. The stochastic nature of wireless channels and various profiles of mobile applications (e.g., task topology and time deadline requirement) present challenges for making decision on application off-loading.

Previous works have provided analysis of off-loading decision for coarse-grained applications under static network condition. Kumar and Lu in [28] presented a rough analysis to decide whether to off-load applications to the cloud, mainly considering computation energy on the mobile device and the communication energy for off-loading over the static network. Miettinen and Nurminen in [35], based on their measurement results, demonstrated the main factors that affect the energy consumption of mobile applications, including workload, data communication patterns, and technologies (i.e., WLAN and 3G). Both the works presented the off-loading policy for the coarse-grained applications under a fixed computation scheduling on the mobile device and a static bandwidth model over wireless networks, which required the prediction of the network condition. Nevertheless, as a mobile application can consist of a set of tasks in the granularity of either method [11] or module [16], it is necessary to consider the fine granularity of mobile applications to reduce the energy consumption on mobile devices. Moreover, due to the varying channel condition, we need to take the stochastic wireless network into account to further reduce the energy consumption on mobile devices.

In this chapter, we focus on the energy-efficient task execution policy in mobile cloud computing. The paradigm of mobile cloud computing provides two basic functionalities for mobile users:

- *Application off-loading*: The cluster of mobile devices [21] or the cloud clone [10] in the infrastructure cloud can execute the application on behalf of mobile devices via Mobile to Mobile (M2M) communication or Mobile to Cloud (M2C) communication, respectively. For M2M communication, off-loading requests are sent via Bluetooth to nearby mobile devices, which is known as ad-hoc virtual cloud. For M2C communication, off-loading requests are sent through base stations or access points to the cloud for the execution. In this chapter, we only consider the infrastructure cloud and its M2C communication for the off-loading policy.

- *Task delegation*: Some applications (e.g., media transcoding and virus scanning) can consume more resources than the clone can afford. In this case, we should delegate the tasks to a back-end cloud infrastructure, which has sufficient computing resources. Task delegation can be realized through either RESTful web services or MapReduce.

This paradigm requires policy to decide how to off-load applications and delegate tasks for energy-efficient task execution. The policy serves as a general framework, which can be adapted for real applications.

For application off-loading, our objective is to minimize the energy consumption on the mobile device while meeting the time constraints for the execution of all the tasks under the stochastic wireless channel. Specifically, we consider three cases in terms of the task topology within mobile applications, that is, one node, linear chain, and general topology. For the first case, the whole application is executed either on the mobile device (i.e., mobile execution) or on the cloud (i.e., cloud execution). We obtain a threshold and find an operational region to determine which execution is more energy efficient [56]. For the second case, each task in the linear chain is sequentially executed, with output data as the input of its subsequent task. We propose a collaborative task execution between the mobile device and the cloud. We formulate the collaborative task execution as a constrained shortest path problem and derive a *one-climb* policy, which indicates that the energy-optimal execution only migrates once from the mobile device to the cloud if ever [57]. For the third case, a mobile application consists of fine-grained tasks in general topology. We formulate the collaborative task execution as a delay-constrained workflow scheduling problem. We leverage partial critical path analysis and adopt *one-climb* policy to schedule the task execution [53].

For task delegation, our objective is to minimize the energy consumption in the cloud for transcoding as a service (TaaS). Tasks of video transcoding can be executed locally on mobile devices, or offloaded to a set of service engines in the cloud. The objective is to reduce the energy consumption on both mobile devices and the cloud for executing transcoding tasks. For the mobile device, we find an operational region to determine whether the task should be off-loaded or not [54]. For the cloud, we leverage Lyapunov optimization framework and propose an online algorithm to reduce energy consumption on service engines while achieving the queue stability [55]. The remainder of this chapter is organized as follows. In Section 2.2, we review the off-loading decision in mobile cloud computing. In Section 2.3, we present the policy for application off-loading. In Section 2.4, we present the policy for task delegation. In Section 2.5, we summarize the whole chapter.

2.2 Review of Off-Loading Decision in Mobile Cloud Computing

Off-loading policy in mobile cloud computing has been extensively investigated in recent years. The approaches can be classified into two categories: technical approach and fundamental approach.

The technical approach uses prediction and machine learning technique to make off-loading decisions. X-ray [39] employed linear regression analysis to construct models to accurately predict the cost and benefits of off-loading. Zhang et al. in [59] leveraged Naive Bayesian learning to figure out the weblet configuration for the optimal execution, given device status (such as upload bandwidth, throughput, power level, etc.) and user's preference (such as monetary cost, power consumption, and execution performance). They recommended that the learning algorithms should be lightweight and efficient for the decision of the off-loading policy.

The fundamental approach relies on optimization and graph theory to make off-loading decisions. Integer program formulation is a typical solution to the off-loading policy for fine-grained applications. CloneCloud [10] decided which methods in the application should be migrated for execution. Particularly, they modeled the energy cost as a function of CPU state, display state, and network state. To minimize the energy consumption, the authors formulated an optimization problem as an integer linear program. Similarly, MAUI [11] provided a formulation of integer linear program to optimize the energy savings under application execution time constraint, by deciding at runtime which methods should be remotely executed. Wishbone [38] formulated the program partitioning problem between servers and embedded nodes as an integer linear program, with the objective of minimizing a combination of network bandwidth and CPU consumption. In addition, Chen et al. in [8] aimed to minimize the combination of energy consumption and the latency for each off-loading request from mobile devices to the cloud, and formulated the optimization problem as a linear program. Giurgiu et al. in [16] constructed a consumption graph to represent an application and proposed greedy algorithms to find a cut in the consumption graph for the application execution. The objective was to minimize the interaction between the mobile device and the cloud as well as the amount of exchanged data. However, it did not consider the energy consumption on the mobile device. Huang et al. in [20] presented a dynamic off-loading algorithm based on Lyapunov optimization, which provides a suboptimal solution to save energy on the mobile device while meeting the application deadline. Lin et al. in [33] derived the optimal off-loading policy by maximizing the expected sum on performance and power consumption using dynamic programming. Balakrishnan and Tham in [3] proposed a two-level genetic algorithm and obtained the energy-efficient task mapping and scheduling on a mobile application to decide which partitions of the application should be off-loaded to cloud resources.

In this chapter, we focus on the fundamental approach for designing the off-loading policy of reducing the energy consumption in mobile cloud computing.

2.3 Policy in Application Off-Loading

In this section, we present off-loading policy for application execution in mobile cloud computing. Since a mobile application can consist of a set of tasks, to model the application execution, we construct a directed acyclic graph as follows:

- In the graph, a node represents a fine-grained task and a link represents the data dependency between adjacent tasks.

- A specific cost is associated with the node and the link for the computation and communication, respectively. The cost can be energy consumption or time delay.

- Two dummy nodes are introduced to represent application initiation and termination, respectively.

Thus, by exploring the granularity of mobile applications, we can have the task graph model and formulate the off-loading policy as an optimization problem over the graph to optimize the cost (e.g., minimize the energy consumption on the mobile device while meeting the application delay deadline).

The optimization problem, owing to its combinatorial nature, is NP-hard. By imposing some regularity conditions of the underlying graphic representation, we can reduce the general framework into a few interesting special cases. Specifically, we focus on the following three cases:

- Application as a node: The task graph has only one node (Figure 2.1a), which represents the whole application. The whole application is either executed on the mobile device or off-loaded to the cloud for execution. We will decide which execution, that is, mobile execution or cloud execution, is more energy-efficient under the stochastic wireless channel model in Section 2.3.1.

- Application as a linear chain: The task graph is a linear chain (Figure 2.1b), which renders a dynamic programming approach. The mobile application is represented by a sequence of fine-grained tasks in a linear chain, each of which is executed either

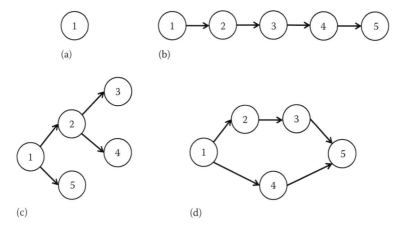

Figure 2.1 Examples of task graphs in different topologies. (a) Node, (b) linear chain, (c) tree, and (d) mesh.

on the mobile device or off-loaded to the cloud for execution. We will provide the energy-efficient execution policy for the application under the stochastic wireless channel model in Section 2.3.2.

- Application as a general topology: The task graph is a general topology (Figure 2.1c and d), which requires a more sophisticated approach. The mobile application is modeled as a more complicated task graph (e.g., tree and mesh). We will design heuristic algorithms to schedule the task execution between the mobile device and the cloud in Section 2.3.3.

We will present the solution of the three cases, respectively.

2.3.1 Application as a Node

In this section, we investigate the optimal application execution policy for the application as a node.

2.3.1.1 Mobile Application Model

We use two parameters to characterize a mobile application, including

- Input data size L: the number of bits as the input data for the application.

- Application completion deadline T_d: the delay deadline before which the application should be completed (Figure 2.2).

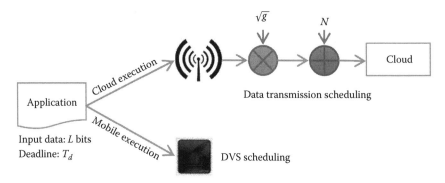

Figure 2.2 A mobile application is either executed on the local mobile device or off-loaded to the cloud for execution.

These two parameters have the impact on the energy consumption of mobile applications. If the mobile application is executed locally (i.e., mobile execution), its energy consumption comes from the computation on the mobile device, which depends on how much workload to be executed within the completion deadline. We will model the workload as a function of input data size. If the mobile application is off-loaded to the cloud for execution (i.e., cloud execution), its energy consumption comes from the data transmission to the cloud, which depends on how many data bits to be transmitted within the completion deadline. Normally, more data input and (or) shorter completion deadline will result in higher energy consumption for the mobile execution and cloud execution. We will use the notation $A(L, T_d)$ to denote the profile of the mobile application, and derive the optimal computation energy of the mobile execution and the optimal transmission energy of the cloud execution, respectively.

2.3.1.2 Optimal Computation Energy of Mobile Execution

When the application is executed on the mobile device, the energy consumption is determined by the computation to accomplish the workload. Specifically, the workload is measured by the number of CPU cycles required by the application. We adopt the workload model [34],

$$W = LX, \tag{2.1}$$

where the workload W depends on the input data size L, and X indicates the number of CPU cycles per bit [34,35]. Particularly, X is a random variable, which reflects the complexity of the algorithm in the application. We denote the probability distribution function (PDF) of X as $p_X(x)$, with its cumulative distribution function (CDF) defined as

$$F_X(x) = \Pr[X \le x], \tag{2.2}$$

and its complementary cumulative distribution function (CCDF) defined as

$$F_X^c(x) = 1 - F_X(x). \tag{2.3}$$

Then, the CDF of the workload W is given by $F_W(w) = F_X(w/L)$, and its CCDF is given by $F_W^c(w) = F_X^c(w/L)$. As shown in [34,50,51], the random variable X can be modeled by a gamma distribution, with the PDF given by

$$p_X(x) = \frac{1}{\theta \Gamma(k)} \left(\frac{x}{\theta}\right)^{k-1} e^{-x/\theta}, \quad \text{for } x > 0, \tag{2.4}$$

depending on the shape parameter k and the scale parameter θ. This workload model is general; we can estimate the gamma distribution according to the characteristics of the specific application. The distribution estimation has been provided in [49–51], and is thus beyond the scope of the chapter.

We adopt a probabilistic performance requirement for workload completion, which requires that each application should meet its deadline within a probability. We denote ρ as the application completion probability (ACP). This parameter ρ is normally set to be very close to one.* Suppose that the allocated CPU cycle is W_ρ. Then, we need to make sure the probability that each application requires no more than the allocated W_ρ cycles should be no less than ρ, that is,

$$F_W(W_\rho) = \Pr[W \le W_\rho] \ge \rho. \tag{2.5}$$

---

* When the application execution fails to meet its deadline, it will continue to execute at the maximum clock frequency for completion. The additional computation energy is negligible since the application completion probability is very close to 1.

The number of CPU cycles, W_ρ, associated with the given ρ, can be given by

$$W_\rho = F_W^{-1}(\rho) = LF_X^{-1}(\rho), \tag{2.6}$$

which is the ρth quantile for the distribution of W.

As in [51], the CPU power consists of dynamic power, short circuit power, and leakage power. Particularly, the dynamic power dominates and hence, we only consider the dynamic power for the mobile execution. In CMOS circuits [7], the energy per operation $\mathcal{E}_w$ is proportional to V^2, where V is the supply voltage to the chip. In addition, the clock frequency of the chip f is approximately linear proportional to the voltage supply V [7]. As a result, the energy per operation can be given by,

$$\mathcal{E}_w(f) = \kappa f^2,$$

where κ is the effective switched capacitance depending on the chip architecture. We set $\kappa = 10^{-17}$ (unit: $\mu J/Hz^2$) so that energy consumption is consistent with the measurements in [35]. The total computation energy for the mobile execution is $\sum_{w=1}^W \mathcal{E}_w(f)$.

For the mobile execution, its total energy consumption can be minimized by optimally configuring the clock frequency of the chip via dynamic voltage scaling (DVS) [40]. Note that we can reduce the energy consumption substantially by setting a low clock frequency for application execution. Due to the delay deadline T_d, however, the clock frequency cannot remain low. As such, we aim to optimally configure the clock frequency to minimize the total energy consumption, while meeting the application delay deadline. Under the optimal CPU frequency scheduling, the minimal energy consumption for the mobile execution is given by,

$$\mathcal{E}_m^* = \min_{\psi \in \Psi}\{\mathcal{E}_m(L, T_d, \psi)\}, \tag{2.7}$$

where
  $\psi = \{f_1, f_2, ..., f_W\}$ is any clock frequency vector under the delay deadline
  Ψ is the set of all feasible clock-frequency vectors
  $\mathcal{E}_m(L, T_d, \psi)$ is the total energy consumed by the mobile device

Specifically, under the probabilistic performance requirement, the optimization problem in Equation 2.7 can be rewritten as,

$$\min_{\{f(w)\}} \quad \kappa \sum_{w=1}^{W_\rho} F_W^c(w)[f(w)]^2, \tag{2.8}$$

$$\text{s.t.} \quad \sum_{w=1}^{W_\rho} \frac{1}{f(w)} \le T_d, \tag{2.9}$$

$$f(w) > 0, \tag{2.10}$$

where Equation 2.8 denotes the total energy consumption during all of these CPU cycles, and Equation 2.9 corresponds to the delay constraint.* This constrained optimization problem can be solved analytically via Largrangian multiplier method and the results are summarized in Theorem 2.1. The derivation of Theorem 2.1 can be found in [56].

Theorem 2.1 *For the mobile execution, the optimal clock scheduling vector is given by*

$$f^*(w) = \frac{\delta}{T_d [F_W^c(w)]^{1/3}}, \quad 1 \le w \le W_\rho, \tag{2.11}$$

where $\delta = \sum_{w=1}^{W_\rho} [F_W^c(w)]^{1/3}$. The optimal energy consumption is

$$\mathcal{E}_m^* = \frac{\kappa}{T_d^2} \left\{ \sum_{w=1}^{W_\rho} [F_W^c(w)]^{1/3} \right\}^3. \tag{2.12}$$

Based on Theorem 2.1, we can characterize the mobile execution by the following propositions.

Proposition 2.1 *The optimal clock frequency increases monotonically with the number of CPU cycles completed. That is, as w becomes larger, the corresponding clock frequency $f(w)$ is larger.*

---

* For each CPU cycle, its execution time is $1/f(w)$, and the energy consumption is $\kappa F_W^c(w)[f(w)]^2$, where $F_W^c(w)$ is the probability that the application has not completed after w CPU cycles.

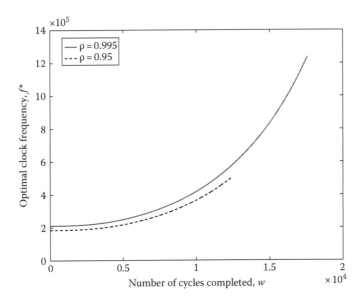

Figure 2.3   The optimal clock frequency, $f^*(w)$, is plotted as a function of the number of CPU cycles w completed.

Example 2.1 *We can illustrate an example for Proposition 2.1. In Figure 2.3, we set $k = 4$, $\theta = 200$, and $T_d = 50\,ms$ and plot the optimal clock frequency as a function of the number of CPU cycles completed. The optimal clock frequency increases monotonically as the number of CPU cycles completed increases. Moreover, if the ACP, ρ, becomes larger, the optimal clock frequency is set to be higher to meet the probabilistic delay deadline.*

Proposition 2.2 *As the application completion probability increases to 1, the minimum energy consumption will converge monotonically to a finite value.*

Example 2.2 *We can illustrate an example for Proposition 2.2. We set $k = 4$, $\theta = 200$, and $T_d = 50\,ms$, and plot in Figure 2.4 the minimum computation energy, $\mathcal{E}_m^*$, as a function of the application completion probability, ρ. We set the input data size L as 800 bits and use this same input data size for the cloud execution in other examples. Notice that the gamma distribution is exponentially tailed. As a result, as the application completion probability of ρ increases, the minimum computation energy increases monotonically and converges to a finite value.*

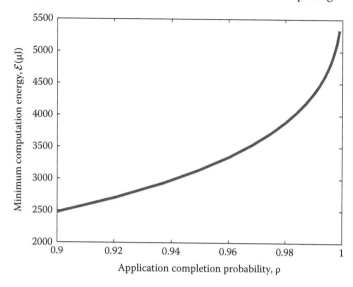

Figure 2.4   The minimum computation energy, $\mathcal{E}_m^*$, is plotted as a function of the application completion probability ρ.

Proposition 2.3 *Under the optimal CPU scheduling, the minimum computation energy is proportional to the negative quadratic of the delay deadline. That is, $\mathcal{E}_m^* \sim T_d^{-2}$.*

Proposition 2.3 can be derived from Equation 2.12.

Proposition 2.4 *Under the optimal CPU scheduling, the minimum computation energy is proportional to cube of the input data size. That is, $\mathcal{E}_m^* \sim L^3$.*

Example 2.3 *We can illustrate an example for Proposition 2.4. We set $k = 4$, $\theta = 200$, and $T_d = 50\,ms$, and plot in Figure 2.5 the minimum computation energy as a function of the input data size. We compare it with a scaling law of L^3. It shows that $\mathcal{E}_m^*$ scales at L^3.*

Based on Propositions 2.3 and 2.4, we can rewrite the optimal energy consumption of mobile execution as

$$\mathcal{E}_m^* = \frac{ML^3}{T_d^2}, \tag{2.13}$$

where M is a constant, depending on parameters κ and ρ for the mobile execution.

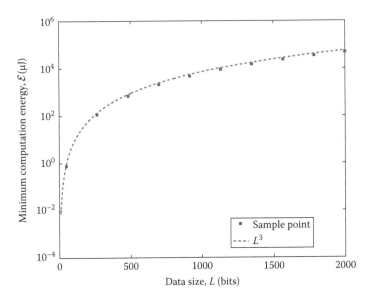

Figure 2.5   The minimum computation energy, $\mathcal{E}_m^*$, is plotted as a function of the input data size L.

2.3.1.3   Optimal Transmission Energy of Cloud Execution

When the application is executed on the cloud, the energy consumption is determined by the communication, depending on the amount of data to be transmitted from the mobile device to the cloud clone and the wireless channel model. Particularly, we consider a discrete time model, under which L bits of input data should be transmitted with a deadline in T_d time slots.

We make the following assumptions for the cloud execution. First, we assume that the binary executable file for the application has been replicated on the cloud clone such that it does not incur additional energy cost for transmitting the executable file. Second, we assume a stochastic wireless fading channel model between the mobile device and the cloud clone, which is characterized by a channel gain and a noise power. Third, the receiving power is a constant [35] and we do not consider the scheduling of the output results from the cloud.

We adopt the transmission energy model proposed in [31,52]. Specifically, the energy consumption for transmitting r bits of data over the wireless channel within a time slot can be approximated by a convex monomial function,

$$\mathcal{E}_t(r,g) = \sigma \frac{r^n}{g}, \tag{2.14}$$

where
  σ denotes the energy coefficient
  n denotes the monomial order
  g denotes the gain of a wireless fading channel

By choosing an appropriate coefficient σ and order n, the resulted monomial function can be close to the capacity-based power model [36,52]. Normally, n takes the value between 2 and 5, that is, $2 \leq n \leq 5$, depending on the modulation scheme. In addition, we set $\sigma = 1.5$ (unit: $\mu J/bit^2$) such that the resulted energy consumption is consistent with the measurements in [35]. Under this energy model, the total energy consumption on the mobile device for cloud execution is $\sum_{t=1}^{T_d} \mathcal{E}_t(r_t, g_t)$, where r_t and g_t are the number of bits transmitted and channel state in time slot t, respectively.

We adopt Gilbert–Elliott (GE) channel model for the stochastic wireless channel in the cloud execution. The channel gain g_t is determined by a discrete state space Markov model [25,52].

In the GE channel model, there are two states, that is, "good" and "bad" channel conditions (denoted as G and B). The two states correspond to a two-level quantization of the channel gain. If the channel gain is above some value, the channel is labeled as good. Otherwise, the channel is labeled as bad. Let the (average) channel gains of the good and bad states be g_G and g_B, respectively. The transition matrix of the channel state is

$$\mathbb{P} = \begin{pmatrix} p_{GG} & p_{GB} \\ p_{BG} & p_{BB} \end{pmatrix},$$

where

p_{GG} denotes the probability that the next state is the good state
if the current state is the good state

p_{BB} denotes the probability that the next state is the bad state
if the current state is the bad state

p_{GB} denotes the probability that the next state is the bad state
if the current state is the good state

p_{BG} denotes the probability that the next state is the good state
if the current state is the bad state

We have $p_{GB} = 1 - p_{GG}$ and $p_{BG} = 1 - p_{BB}$. Since the state sojourn time is geometrically distributed, the number of time slots being in the good and bad state is $T_G = 1/(1 - p_{GG})$ and $T_B = 1/(1 - p_{BB})$, respectively.

For the cloud execution, its total energy consumption can be minimized by optimally varying the number of bits transmitted in the time slot in response to a stochastic channel. Since the energy cost per time slot increases with the number of bits to be transmitted, we would transmit as few bits as possible [42]. However, reducing the number of bits transmitted per time slot could violate the delay constraint of the application execution. We aim to find out an optimal transmission data-rate schedule to minimize the total transmission energy, while satisfying the delay constraint. Under the optimal transmission scheduling, the minimal transmission energy for the cloud execution is given by

$$\mathcal{E}_c^* = \min_{\phi \in \Phi} \mathbb{E}\{\mathcal{E}_c(L, T_d, \phi)\}, \tag{2.15}$$

where

$\phi = \{r_1, r_2, ..., r_{T_d}\}$ denotes a data transmission vector under the delay deadline

Φ is the set of all feasible data transmission vectors

$\mathcal{E}_c(L, T_d, \phi)$ denotes the transmission energy with the expectation taken for different channel states

We denote t as discrete time index in descending order (i.e., from T_d to 1). Then, the optimization problem in Equation 2.15 for the optimal data-transmission schedule can be rewritten as,

$$\min_{r_t} : \quad \mathbb{E}\left[\sum_{t=1}^{T_d} \mathcal{E}_t(r_t, g_t)\right]$$

$$\text{s.t.:} \quad \sum_{t=1}^{T_d} r_t = L, \tag{2.16}$$

$$r_t \geq 0, \forall t.$$

The minimum expected energy depends on the channel state at $t = T_d + 1$ ($g_{T_d+1} = G$ or $g_{T_d+1} = B$). We denote the condition $g_{T_d+1} = G$ and $g_{T_d+1} = B$ as $(.;G)$ and $(.;B)$, respectively, and find the optimal number of data bits transmitted in each time slot r_t^* and the minimum energy $\mathcal{E}_c^*$ under the two condition. Specifically, the optimal data transmission for the cloud execution is provided in Theorem 2.2. The derivation of Theorem 2.2 can be found in [56].

> In practice, r_t is an integer and Equation 2.16 becomes an integer programming problem, which could be hard to solve. To gain analytical insight, we assume that r_t is continuous in order to obtain the closed-form solution, which provides a lower bound to the actual optimization problem.

Theorem 2.2 *Denote l_t as the number of unfinished bits at time slot t. Under the cloud execution, the optimal data transmission vector is*

$$r_t^*(l_t, g_t; G) = \begin{cases} l_t\left[\dfrac{(g_t)^{1/(n-1)}}{(g_t)^{1/(n-1)}+(\frac{1}{\zeta_{t-1;G}})^{1/(n-1)}}\right], & t \geq 2, \\ l_1, & t = 1, \end{cases} \tag{2.17}$$

if $g_{T_d+1} = G$, where

$$\zeta_{t;G} = \begin{cases} p_{GG}\left[\left(\dfrac{1}{(g_G)^{1/(n-1)}+(\frac{1}{\zeta_{t-1;G}})^{1/(n-1)}}\right)^{n-1}\right] \\ \quad + p_{GB}\left[\left(\dfrac{1}{(g_B)^{1/(n-1)}+(\frac{1}{\zeta_{t-1;G}})^{1/(n-1)}}\right)^{n-1}\right], & t \geq 2, \\ p_{GG}\left[\frac{1}{g_G}\right]+p_{GB}\left[\frac{1}{g_B}\right], & t = 1; \end{cases} \tag{2.18}$$

and

$$r_t^*(l_t, g_t; B) = \begin{cases} l_t\left(\dfrac{(g_t)^{1/(n-1)}}{(g_t)^{1/(n-1)}+(\frac{1}{\zeta_{t-1;B}})^{1/(n-1)}}\right), & t \geq 2, \\ l_1, & t = 1, \end{cases}$$ (2.19)

if $g_{T_d+1} = B$, where

$$\zeta_{t;B} = \begin{cases} p_{BB}\left[\left(\dfrac{1}{(g_B)^{1/(n-1)}+\left(1/\zeta_{t-1;B}\right)^{1/(n-1)}}\right)^{n-1}\right] \\ \quad + p_{BG}\left[\left(\dfrac{1}{(g_G)^{1/(n-1)}+\left(1/\zeta_{t-1;B}\right)^{1/(n-1)}}\right)^{n-1}\right], & t \geq 2, \\ p_{BB}\left[\dfrac{1}{g_B}\right] + p_{BG}\left[\dfrac{1}{g_G}\right], & t = 1. \end{cases}$$ (2.20)

Correspondingly, the minimum transmission energy is

$$\mathcal{E}_c^*(L, T_d; G) = \sigma L^n \zeta_{T_d;G},$$ (2.21)

and

$$\mathcal{E}_c^*(L, T_d; B) = \sigma L^n \zeta_{T_d;B},$$ (2.22)

respectively. Therefore, the minimum expected transmission energy $\mathcal{E}_c^*$ is

$$\mathcal{E}_c^*(L, T_d) = \frac{T_G}{T_G + T_B}\mathcal{E}_c^*(L, T_d; G) + \frac{T_B}{T_G + T_B}\mathcal{E}_c^*(L, T_d; B).$$ (2.23)

Based on Theorem 2.2, we can have the following propositions.

Proposition 2.5 *As the input data size L increases, the minimum transmission energy $\mathcal{E}_c^*$ increases monotonically and scales with a factor of L^n, where n is the monomial order. That is, $\mathcal{E}_c^* \sim L^n$.*

Proposition 2.6 *As the application completion deadline T_d increases, the minimum transmission energy $\mathcal{E}_c^*$ decreases monotonically and scales with a factor of $T_d^{-(n-1)}$, where n is the monomial order. That is, $\mathcal{E}_c^* \sim T_d^{-(n-1)}$.*

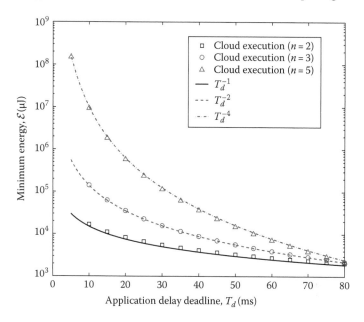

Figure 2.6   Expected transmission energy is plotted as a function of deadline T_d for the GE channel model.

Example 2.4 *We can illustrate an example for Proposition 2.6. We set $L = 800\,bits$, $p_{GG} = 0.995$, $p_{BB} = 0.96$, $g_G = 1$, and $g_B = 0.1$, and plot in Figure 2.6 the expected transmission energy under GE model as a function of the deadline T_d with different n. We compare them with a scaling factor of $T_d^{-(n-1)}$. Note that, the scaling factor matches the numerical results well, which is consistent with Proposition 2.6. Moreover, as the application delay deadline T_d becomes smaller, the term $T_d^{-(n-1)}$ will be larger, which results in more energy consumption.*

Based on Propositions 2.5 and 2.6, we can rewrite the optimal energy consumption of the cloud execution as

$$\mathcal{E}_c^* = \frac{C_0(n)L^n}{T_d^{n-1}}, \tag{2.24}$$

where $C_0(n)$ depends on the monomial order n, the energy coefficient σ, the channel states (good or bad), and the state transition probability of GE model.

2.3.1.4   Optimal Application Execution Policy

In previous sections, we have obtained the optimal energy consumption for the mobile execution and the cloud execution, respectively. To achieve the optimal application execution, we can compare the minimum computation energy of mobile execution and the minimum transmission energy of cloud execution. The optimal application execution policy is to choose whichever consumes less energy on the mobile device. As such, we can adopt the following decision rule,

$$\begin{cases} \text{Mobile execution} & \text{if} \quad \mathcal{E}_m^* \le \mathcal{E}_c^* \\ \text{Cloud execution} & \text{if} \quad \mathcal{E}_m^* > \mathcal{E}_c^*. \end{cases} \tag{2.25}$$

in order to minimize the total energy consumed on the mobile device.

Example 2.5 *We use the same application parameters of the mobile execution and the cloud execution as those described previously. That is, the application workload is modeled as the gamma distribution, with $k = 4$, $\theta = 200$, $L = 800\,bits$. The channel is a GE model with $p_{GG} = 0.995$, $p_{BB} = 0.96$, $g_G = 1$, and $g_B = 0.1$. In Figure 2.7, we plot the minimum energy consumed by the mobile device for the mobile execution and the*

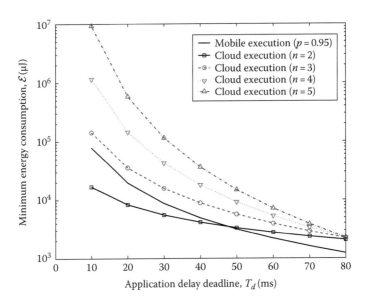

Figure 2.7   The minimum energy, $\mathcal{E}^*$, is plotted as a function of the application delay deadline T_d.

cloud execution, as a function of the application completion deadline T_d. If n is smaller than 3 (e.g., n = 2), the cloud execution is more energy-efficient when the delay deadline is below a threshold. However, if n is larger than 3 (e.g., n = 3, 4, 5), the mobile execution is more energy-efficient. Hence, under a fixed input data size L, the monomial order of n can affect the optimal execution decision. Moreover, by optimally deciding where to execute the application, a significant amount of energy can be saved on the mobile devices. For example, for n = 2, when the delay deadline is 10 ms, the cloud execution consumes energy 4.65 times less than the mobile execution.

Example 2.6 *We set k = 4, θ = 200, T_d = 50 ms, p_{GG} = 0.995, p_{BB} = 0.96, g_G = 1, and g_B = 0.1. In Figure 2.8, we plot the minimum energy consumption on the mobile device for the mobile execution and the cloud execution, as a function of the input data size of L. If n is smaller than 3, the cloud execution is more energy-efficient when the data size is beyond a threshold. However, if n is larger than 3, the mobile execution is more energy-efficient. Hence, under a fixed delay deadline, the monomial order of n can affect the optimal execution decision.*

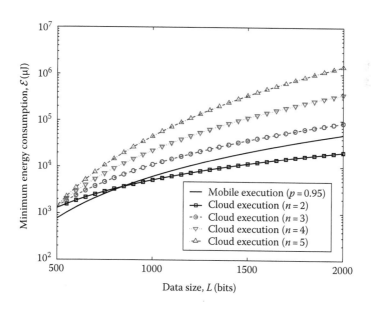

Figure 2.8   The minimum energy, $\mathcal{E}^*$, is plotted as a function of the data size L.

We develop a threshold policy for the optimal application execution, based on the analytical results obtained in the mobile execution and the cloud execution. We define the effective data consumption rate as the ratio of input data size and delay deadline ($R_e = L/T_d$), and a threshold $R_{th} = [M/C_0(n)]^{1/(n-3)}$. Specifically, the effective data consumption rate R_e refers to the data processing rate for the mobile execution, or the data transmission rate for the cloud execution. R_{th} can be regarded as the slope of the line geometrically. Since R_{th} depends on the monomial order (n), we can have different decisions for the energy-optimal application execution, which is given by Theorem 2.3.

Theorem 2.3 *The energy-optimal execution policy can be obtained by comparing the effective data consumption rate ($R_e = L/T_d$) with a threshold $\left(R_{th} = [M/C_0(n)]^{1/(n-3)}\right)$. Specifically, the execution decision for mobile device is as follows:*

$$(1)\ n < 3 \quad \begin{matrix} \text{mobile execution} & \text{if} & R_e \leq R_{th} \\ \text{cloud execution} & \text{if} & R_e > R_{th} \end{matrix} ; \qquad (2.26)$$

$$(2)\ n = 3 \quad \begin{matrix} \text{mobile execution} & \text{if} & M \leq C_0(n) \\ \text{cloud execution} & \text{if} & M > C_0(n) \end{matrix} ; \qquad (2.27)$$

$$(3)\ n > 3 \quad \begin{matrix} \text{mobile execution} & \text{if} & R_e \geq R_{th} \\ \text{cloud execution} & \text{if} & R_e < R_{th} \end{matrix} . \qquad (2.28)$$

Theorem 2.3 demonstrates that the off-loading policy depends on the application profile (i.e., the input data size L and the completion deadline T_d), the wireless transmission model (i.e., the monomial order n), and parameters M and $C_0(n)$ (depending on effective switched capacitance κ on the chip system of the device and energy coefficient σ in the wireless channel model). Therefore, we can obtain the energy-optimal execution policy by simply computing the effective data consumption rate (L/T_d).

Example 2.7 *For various application profiles $A(L, T_d)$ with L ranging from 0 to 1000 bits and T_d ranging from 0 to 100 ms, we compute the energy consumption of mobile execution and cloud execution using the same parameters in Figures 2.7 and 2.8, and plot in Figure 2.9 the regions where the mobile execution or the cloud execution is more energy-efficient under different n. Specifically, for $n = 2$, the two optimal operational regions are separated by a line. That is, if the point (T_d, L) is above the line, the cloud execution is more energy-efficient; otherwise, the mobile execution is more energy-efficient. With the same parameters for both mobile*

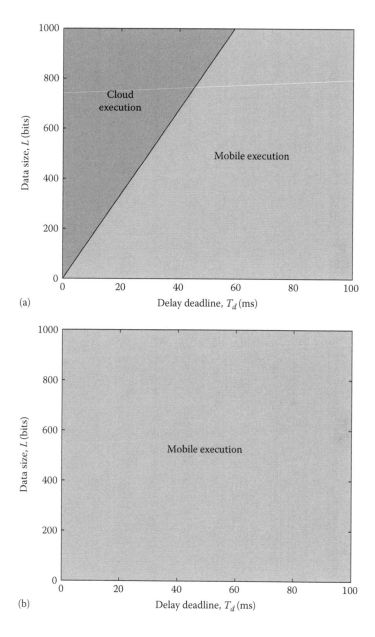

Figure 2.9 Operational regions for optimal energy decision. $\sigma = 1.5$. (a) $n = 2$ and (b) $n = 3, 4, 5$.

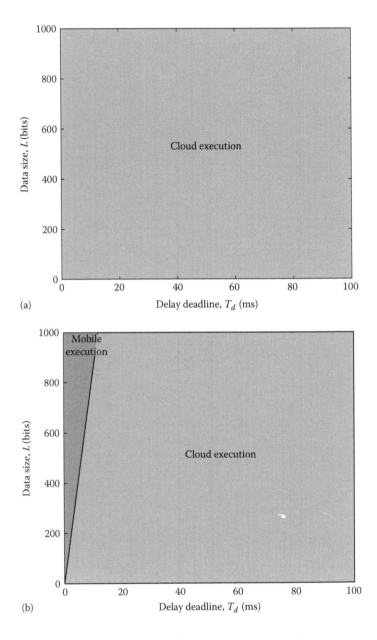

Figure 2.10 Operational regions for optimal energy decision. σ = 0.1. (a) *n* = 2, 3, (b) *n* = 4.   (*Continued*)

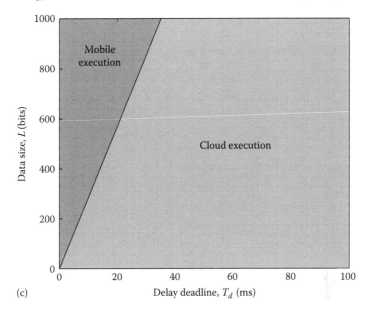

(c)

Figure 2.10 (*Continued*) Operational regions for optimal energy decision. $\sigma = 0.1$. (c) $n = 5$.

execution and cloud execution, all cases of $n = 3, 4, 5$ should be executed in the mobile device. This can be derived from Figures 2.7 and 2.8, in which for $n = 3, 4, 5$, the curve of the cloud execution is always above the curve of the mobile execution, indicating that energy consumption of mobile execution is smaller. Thus, the mobile execution is more energy-efficient.

Example 2.8 *We adopt the same application parameters as in the previous example except that we set $\sigma = 0.1$ and investigate its impact to the optimal operational regions. The same approach can be adopted to obtain the optimal policy for application execution. By reducing the value of σ, the energy consumption of cloud execution is reduced. In Figure 2.10, for $n = 2, 3$, the cloud execution is more energy-efficient than the mobile execution. While for $n = 4, 5$, there exists a line to separate the regions of the cloud execution and the mobile execution.*

2.3.2   Application as a Linear Chain

In this section, we investigate the energy-efficient off-loading policy for the application as a linear chain.

2.3.2.1   System Models and Problem Formulation

Task Model

Figure 2.11 illustrates the task model in the linear chain. There are n tasks in the mobile application. Each task is sequentially executed, with output data as the input of its subsequent task. We define a tuple notation to represent the context of each task, $\phi_k = (\omega_k, \alpha_k, \beta_k)$, where $k = 1, 2, ..., n$. Specifically, ω_k is the computing workload of the kth task. As such, the total computing workload of the application is $\sum_{k=1}^{n} \omega_k$. We denote the input and output data size of the kth task as α_k and β_k, respectively. Notice that the output of the $(k-1)$th task is the input of the kth task, that is, $\alpha_k = \beta_{k-1}$.

Example 2.9 *Eyedentify [27], a color-based object recognition, is a typical application example of the linear chain. For feature extraction process, EyeDentify consists of a series of steps to convert the raw image into a feature vector. Given an image, we indicate a number of circular areas around the center of the image, and establish color histograms for each of the circular areas. Then, we process the shape of the color histograms toward a weibull fit. Combining all of the fits for all the circular areas, we obtain a feature vector, which represents the description of the image. Finally, we can accomplish object recognition by selecting the best match for the feature vector from the local database of objects.*

Wireless Channel Model

Wireless channel is modeled as a random process of the channel state g_t at time slot t under a discrete time slot scheme. Specifically, we consider three models, including

- Block-fading channel, where $\{g_t\}$ are fixed and remain the same over the duration of application execution.

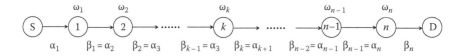

Figure 2.11   Illustration of task model in a linear chain. The label above a node denotes the corresponding workload of the task, while the label below denotes the input/output data.

- IID stochastic channel, where $\{g_t\}$ are independent and identically distributed (i.i.d.).

- Markovian stochastic channel, where $\{g_t\}$ form a Markovian random process under GE channel model.

We assume that the transmission power on the mobile device is fixed. As a result, the data rate R_t is fully determined by the channel state of g_t.

Under the Markovian stochastic channel, we assume that the data rate takes two values (R_G or R_B) corresponding to the channel states (g_G or g_B), that is,

$$R_t = \begin{cases} R_G, & g_t = g_G; \\ R_B, & g_t = g_B. \end{cases}$$

Execution Model

We consider the collaborative task execution between the mobile device and the cloud clone. Since there can be interactions between the mobile device and the cloud, four atomic modules will be involved during the collaborative task execution, including

- *Mobile Execution (ME)*: The task is executed on the mobile device. The completion time for the mobile execution depends on the workload of the task and the clock frequency of the mobile device. Under the mobile execution, the completion time of the kth task is

$$d_m(k) = \omega_k f_m^{-1}, \tag{2.29}$$

where f_m is the clock frequency of the mobile device. We assume that f_m is fixed.* Correspondingly, the energy consumption of the kth task on the mobile device is

$$e_m(k) = d_m(k)p_m, \tag{2.30}$$

where p_m is the computation power of the mobile device. We assume that p_m is fixed and does not change during the computation.

---

* If the clock frequency is dynamic, we will need to determine not only whether a task should be off-loaded for the cloud execution, but also the clock frequency on the mobile device for the mobile execution. Then, the problem will become unnecessarily complex in the linear chain case, and thus we set the clock frequency of the mobile device to be fixed.

- *Cloud Execution (CE)*: The task is executed on the cloud clone. The completion time for the cloud execution depends on the workload of the task and clock frequency of the cloud clone. Under the cloud execution, the mobile device becomes idle and wireless network interface card is turned off. Then, the completion time of the kth task is

$$d_c(k) = w_k f_c^{-1}, \qquad (2.31)$$

where f_c is the clock frequency of the cloud clone, which is faster than the CPU clock frequency of the mobile device, that is, $f_c > f_m$. In this case, we have $d_c(k) < d_m(k)$. Correspondingly, the energy consumption of the kth task on the mobile device is

$$e_c(k) = d_c(k) p_0, \qquad (2.32)$$

where p_0 is the power of the mobile device for being idle. We assume that $p_0 < p_m$ and hence $e_c(k) < e_m(k)$.

- *Sending Input Data (SID)*: Input data of the task is sent to the cloud clone. If the $(k-1)$th task is executed on the mobile device while the kth task is off-loaded to the cloud for execution, the input data of the k task, α_k, should be sent to the cloud before execution. We denote $d_s(k)$ as the transmission time for the kth task, and suppose the current time slot is i, then we have

$$d_s(k) = \min\left\{ j : \sum_{t=i}^{i+j-1} R_t \ge \alpha_k \right\}. \qquad (2.33)$$

Correspondingly, the energy consumption of the mobile device is

$$e_s(k) = d_s(k) p_s, \qquad (2.34)$$

where p_s is the transmission power of the mobile device.

- *Receiving Output Data (ROD)*: Output data of the task is received by the mobile device. If the $(k+1)$th task is executed on the mobile device while the kth task is executed on the cloud, the output data of the kth task, β_k, should be received by the mobile device before the commencement of executing the $(k+1)$th task. We denote $d_r(k)$

as the completion time of receiving the output data of the kth task, and suppose the current time slot is i, then we have

$$d_r(k) = \min\left\{j : \sum_{t=i}^{i+j-1} R_t \geq \beta_k\right\}. \qquad (2.35)$$

Correspondingly, the energy consumption of the mobile device is

$$e_r(k) = d_r(k)p_r, \qquad (2.36)$$

where p_r is the receiving power of the mobile device.

Graph Representation and Problem Formulation

To represent the collaborative task execution between the mobile device and the cloud, we construct a directed acyclic graph $G = (V, A)$, with the finite node set V and arc set A, shown in Figure 2.12. Two dummy nodes, that is, node S and node D, are added as the source node for application initiation and the destination node for application termination, respectively. The node k represents that the kth task has been completed on the mobile device, while the node $\bar{k}$ represents that the kth task has been completed by the cloud clone, where $k = 1, 2, ..., n$.

The arc of two nodes is associated with a nonnegative cost C generalized for the energy consumption and the completion time, bracketed with the module involved. The cost depends on the execution decision on the two corresponding tasks:

- From node S to node 1, the module ME is involved if task 1 is executed on the mobile device, or the module SID is involved, followed by the module CE if it is off-loaded to the cloud.

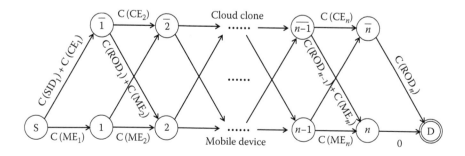

Figure 2.12 Representation of the task execution flow.

- From node k to node $\overline{k+1}$, the module SID will take place followed by the module CE, with the positive cost of $e_{k,\overline{k+1}}$ and $d_{k,\overline{k+1}}$ for the energy consumption and time delay, respectively.

- From node $\overline{k}$ to node $k+1$, the module ROD will take place followed by the module ME, with the energy consumption $e_{\overline{k},k+1}$ and time delay $d_{\overline{k},k+1}$, respectively.

- The cost between n and D is zero, while the cost between $\overline{n}$ and D is incurred by the module ROD.

With the graph representation, we can transform the collaborative task execution as a constrained optimization problem in finding the shortest path in terms of expected energy consumption between S and D, under the constraint that the expected completion time of that path should be no more than time deadline. A path p is feasible if its time delay satisfies the constraint. A feasible path p^* with the minimum expected energy consumption is the optimal solution. Mathematically, we have

$$
\begin{aligned}
\min_{p \in \mathcal{P}} \quad & \mathbb{E}[e(p)] = \mathbb{E}\left\{ \sum_{(u,v) \in p} e_{u,v} \right\} \\
\text{s.t.} \quad & \mathbb{E}[d(p)] = \mathbb{E}\left\{ \sum_{(u,v) \in p} d_{u,v} \right\} \le T_d,
\end{aligned}
\tag{2.37}
$$

where
  T_d is the completion deadline of the entire application
  $\mathcal{P}$ is the set of all feasible paths, and the expectation is taken over the channel state

This constrained optimization problem is NP-complete [46], because there are 2^n possible options for the solution as we have two choices for each task, that is, off-loading to the cloud or not.

2.3.2.2 Characterization of Optimal Solution

To solve Equation 2.37, we first consider a simple case under the block-fading channel [5], with a constant data rate R during the execution of all the tasks of the application. This simple case can give some guidelines for the design of the scheduling policy.

Under the block-fading channel, the problem of minimum energy consumption within time delay can be transformed into a deterministic constrained shortest path problem. Specifically, all the costs of data transmission are deterministic, including the completion time and the energy consumption of each task, that is, $d_s(k) = \alpha_k/R$, $d_r(k) = \beta_k/R$, and $e_s(k) = (\alpha_k/R)p_s$, $e_r(k) = (\beta_k/R)p_r$. In this case, one can enumerate all the paths and choose the one that results in the minimal energy consumption while satisfying the time delay constraint. However, this brute-force search will lead to the complexity of $O(2^n)$. Hence, an efficient algorithm is needed to reduce the complexity.

Before presenting the efficient algorithm, we first characterize the optimal solution and show that the *one-climb* policy is optimal for the block-fading channel in Theorem 2.4. The derivation of Theorem 2.4 can be found in [57].

Theorem 2.4 *Under the block-fading channel model, the energy-optimal execution for the linear chain task topology only migrates once from the mobile device to the cloud if ever. We refer to such an execution as one-climb policy.*

By the rule of the *one-climb* policy, there will occur at most one task migration for the optimal policy during the entire application execution. For example, the execution in Figure 2.13a is not optimal, because, the execution in Figure 2.13b can have less energy consumption and execution time.

Based on the *one-climb* policy, we can design an efficient algorithm for task scheduling, by enumerating all the paths under the *one-climb* policy. We denote $\mathcal{P}'$ as the set of all the paths under the *one-climb* policy. There are $((n+1)n/2) + 1$ paths in $\mathcal{P}'$, thus the searching space of the *one-climb* policy is greatly reduced compared to the brute-force search. For each *one-climb* path, we compute the energy consumption and delay by summing the weights of arcs along the path. After computing the cost of all the paths in $\mathcal{P}'$, we can choose the one with the minimum energy consumption while the time delay is within the deadline. The details are illustrated in Algorithm 2.1. The complexity of Algorithm 2.1 is $O(n^3)$.

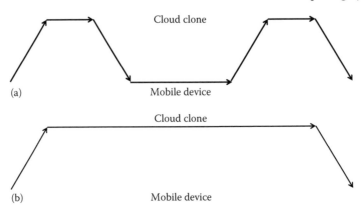

Figure 2.13 Examples of task execution for linear chain case between the mobile device and the cloud clone. (a) Execution with two migrations from the mobile device to the cloud clone. (b) Execution with one migration from the mobile device to the cloud clone.

Algorithm 2.1 The enumeration algorithm based on *one-climb* policy

---

Require: $\mathcal{P}'$
Ensure: $p^*, e(p^*)$
  Set $e_0 = M_E$, where M_E is a very large value
  for $p \in \mathcal{P}'$ **do**
    if $d(p) \le T_d$ and $e(p) < e_0$ **then**
      $p_0 = p$
      $e_0 = e(p)$
    end if
  end for
  if $e_0 == M_E$ **then**
    return "There is no solution."
  else
    $p^* = p_0$
    $e(p^*) = e_0$
  end if

We can transform the optimization problem under the stochastic network into the deterministic shortest path problem. Under the IID stochastic channel, we need to find the expected cost for the data transmission. The data transmission can be formulated as a stopping-time problem, that is, when the data will be completely transmitted. According to Wald's Equation [44], we can have

$$\mathbb{E}\left\{\sum_{t=1}^{\iota} R_t\right\} = \mathbb{E}(\iota)\mathbb{E}(R), \tag{2.38}$$

where ι is the stopping time to transmit the data. As such, we can approximate the expected time of sending the input data and receiving the output data for each task under the IID stochastic channel. Following that, we can apply the enumeration algorithm based on the *one-climb* policy to find out the path with the minimum energy consumption while satisfying the time delay constraint.

2.3.2.3 Approximate Task Execution Algorithm

We can also leverage LARAC algorithm to find the approximate solution for the constrained shortest path problem and design the task execution algorithm [26], the complexity of which is $O(n^2 \log^2 n)$. The optimization problem in Equation (2.37) is equivalent to the following optimization problem,

$$\min\{\mathbb{E}[e(p)]|p \in P(S,D), \mathbb{E}[d(p)] \le T_d\}, \tag{2.39}$$

where $P(S,D)$ is the set of paths from S to D. We can define a Lagrangian function

$$L(\lambda) = \min\{\mathbb{E}[e_\lambda(p)]|p \in P(S,D)\} - \lambda T_d, \tag{2.40}$$

where $\mathbb{E}[e_\lambda(p)]$ is the aggregated cost given by

$$\mathbb{E}[e_\lambda(p)] = \mathbb{E}[e(p)] + \lambda\mathbb{E}[d(p)], \tag{2.41}$$

and λ is the Lagrangian multiplier that denotes a cost penalty per time unit if the total completion time exceeds the given time delay.

LARAC, Lagrange relaxation based aggregated cost, was proposed in [26] to solve the delay constrained least cost path problem. The LARAC algorithm has the following properties:

- It can return a path that satisfies the delay constraint.

- It provides a lower bound of the optimal solution (i.e., $L(\lambda) \le \mathbb{E}[e(p^*)]$).

- Its running time is polynomial.

Algorithm 2.2 illustrates how we can leverage LARAC algorithm to find a path with the minimum expected aggregated cost under the delay deadline. Particularly, **ShortestPath** $(S, D, .)$ is a key procedure that finds the shortest path between the source node S and the destination node D, where the third input parameter denotes a cost metric, that is, energy consumption (e), time delay (d), and aggregated cost (e_λ). If the shortest path in terms of energy can meet the deadline, then we can obtain the optimal solution. If the shortest path in terms of time cannot meet the deadline, then there is no solution; if both the two conditions are not satisfied, we iteratively update p_e and p_d to find the optimal λ for the collaborative task execution.

For the block-fading channel and IID stochastic channel model, we can apply backward induction using Bellman equation to compute

---

Algorithm 2.2 Algorithm based on LARAC for collaborative task execution

---

Require: $G(\mathcal{V}, \mathcal{A})$
Ensure: p_λ^*
  $p_e = $ **ShortestPath**(S, D, e)
  if $d(p_e) \le T_d$ **then**
    return p_e
  end if
  $p_d = $ **ShortestPath**(S, D, d)
  if $d(p_d) > T_d$ **then**
    return "There is no solution"
  end if
  while true **do**
    $\lambda = \frac{e(p_e) - e(p_d)}{d(p_d) - d(p_e)}$
    $p_\lambda = $ **ShortestPath**(S, D, e_λ)
    if $e_\lambda(p_\lambda) = e_\lambda(p_e)$ **then**
      return p_d
    else
      if $d(p_\lambda) \le T_d$ **then**
        $p_d = p_\lambda$
      else
        $p_e = p_\lambda$
      end if
    end if
  end while

---

the minimum energy consumption, time delay, and aggregated cost in Figure 2.12, respectively. Specifically, we first start the calculation from n (and $\bar{n}$) to the node D, and then from $n-1$ (and $\overline{n-1}$) to the node D, until we reach the node S and complete the calculation from S to D.

Example 2.10 *We investigate the performance of the approximate task execution algorithm. Suppose that there are 10 tasks in the application, with time constraint of $T_d = 0.6\,s$. The workload is $\omega = \{40,20,50,30,50,20,40,30,30,20\}M$ cycles. The input data is $\alpha = \{10,4,0.1,0.2,0.1,0.2,0.1,0.2,0.1,0.1\}$ kb. The output data is $\beta = \{4,0.1,0.2,0.1,0.2,0.1,0.2,0.1,0.1,10\}$ kb. We set the parameters of the mobile device and the cloud in Table 2.1. We plot in Figure 2.14 the expected energy of the collaborative execution under the time constraint (e.g., 0.6 s) for various data rates. It shows that the minimum expected energy consumption is a piecewise function of the data rate. Specifically, the execution policy is $\{0,0,1,1,1,1,1,1,1,0\}$ for 10 kb/s, $\{0,1,1,1,1,1,1,1,1,0\}$ for 20 kb/s, $\{1,1,1,1,1,1,1,1,1,0\}$ for 30 and 40 kb/s, $\{1,1,1,1,1,1,1,1,1,1\}$ for 50–100 kb/s, respectively. As the data rate increases, there are more tasks to be off-loaded to the cloud for execution.*

In addition, we compare the energy consumption using Algorithm 2.2 with the optimal solution. Figure 2.15 shows that Algorithm 2.2 can achieve the optimal solution for most cases except when the data rate is around 20 kb/s, with less than 5% error to the optimal solution. Therefore, Algorithm 2.2 is effective to find the solution to the collaborative execution.

Example 2.11 *We find that the one-climb policy is applicable for the stochastic IID channel model. We randomly generate four application profiles, including the workload and input/output data of the tasks, and then*

Table 2.1 Parameters of Configuration of Mobile Device and Cloud Clone

Power of sending data	$p_s = 0.1$W
Power of receiving data	$p_r = 0.05$W
Power of computing	$p_m = 0.5$W
Power of being idle	$p_0 = 0.001$W
CPU frequency of the mobile device	$f_m = 500$MHz
CPU frequency of the cloud clone	$f_c = 3000$MHz

Advances in Mobile Cloud Computing Systems

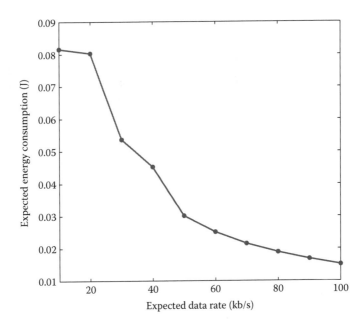

Figure 2.14   Minimum energy consumption as a function of expected data rate.

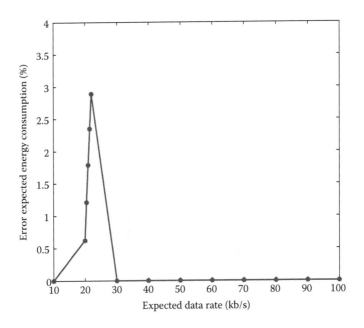

Figure 2.15   Relative error of energy consumption for Algorithm 2.2.

obtain the task scheduling policies. In Figure 2.16, the horizontal axis represents the tasks to be executed, the left vertical axis represents the value of the workload and the input/output data of the tasks, and the right vertical axis represents the optimal task execution location. For a particular task, the darker bar in the middle represents the workload of the task, while its left and right lighter bars represent the input and output data of the task, respectively. The line represents the optimal execution policy for the tasks. It shows that the one-climb *policy holds for all these application profiles.*

For the Markovian stochastic channel model, we can adopt the Markov Decision Process (MDP) and obtain the approximate solution using Algorithm 2.2. Figure 2.17 illustrates the state transition of the collaborative task execution, given that the initial channel state is observed to be good. We construct $n+3$ stages for the task execution. Stage 0 and stage $n+2$ represent the initiation and the termination of application execution, respectively. Stage k ($k = 1, 2, ..., n$) represents that the kth task has been completed. Stage $n+1$ is an intermediate stage before we reach stage $n+2$. In each stage, we define $\mathbf{x_k} = (l_k, g_k)$ as the system state, where l_k is a location indicator that denotes the location where the kth task has been executed (i.e., 0 for mobile device and 1 for cloud clone), and g_k is the channel state of the next time slot we observe when the kth task has been completed. Particularly, we have $l_0 = 0$ and $l_{n+1} = 0$, since the application execution starts on the mobile device and the output results must be resided on the mobile device. In addition, node D is connected to the nodes $(0, G)$ and $(0, B)$ when the final task has been completed, resulting in zero costs for the energy consumption, time delay, and aggregated cost. We also define π_k as the decision variable at stage k that denotes the choice for which the kth task should be executed,

$$\pi_k = \begin{cases} 0, & \text{mobile execution;} \\ 1, & \text{cloud execution.} \end{cases}$$

Taking the decision π_{k+1}, we move from the current state $\mathbf{x_k}$ to the system state $\mathbf{x_{k+1}}$. Since the output of the last task should be resided on the mobile device, we have $\pi_{n+1} = 0$. Under this framework, we can obtain the scheduling policy $\Pi = \{\pi_k\}$, by establishing iterative equations to find the minimum expected time delay, energy consumption, and aggregated cost in Figure 2.17, respectively.

We denote $h_k(\mathbf{x_k})$ as the minimum expected execution time from state $\mathbf{x_k}$ at stage k to state $\mathbf{x_{n+2}}$ at stage $n+2$. As such, $h_0(\mathbf{x_0})$ is the

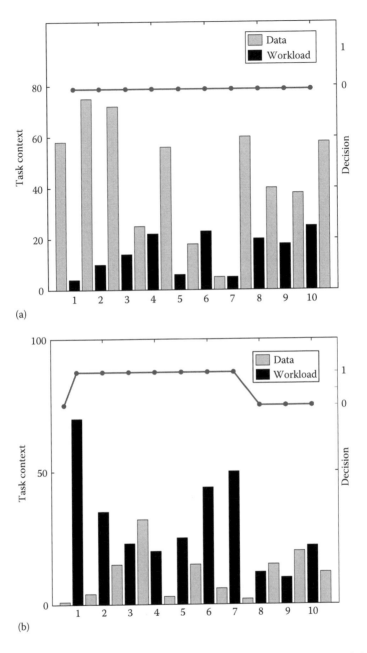

Figure 2.16 Task scheduling policy under the i.i.d. channel model. Time deadline is $T_d = 0.6$s. Workload is in M cycles unit and data is in kb unit. (a and b) $E(R) = 10$ kb/s.   (*Continued*)

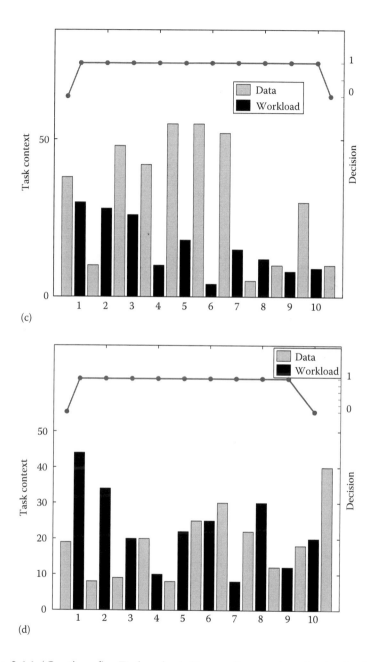

Figure 2.16 (*Continued*) Task scheduling policy under the i.i.d. channel model. Time deadline is $T_d = 0.6$s. Workload is in M cycles unit and data is in kb unit. (c and d) $E(R) = 100$ kb/s.

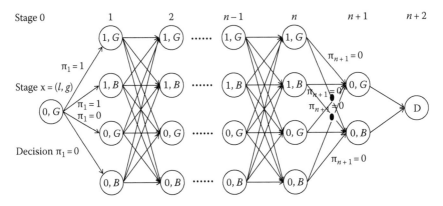

Figure 2.17   Illustration of the state transition of the system. Each state is characterized by the execution location and the channel state. A state transition refers to the transition from one state in a stage to another state in the next stage by applying a decision.

minimum expected time to complete the entire application. Given $h_{k+1}(\mathbf{x_{k+1}})$ for state $\mathbf{x_{k+1}}$ at stage $k+1$, we make the decision for each state at stage k to minimize the expected time delay from state $\mathbf{x_k}$ to state $\mathbf{x_{n+2}}$, based on the backward value iteration,

$$h_k(\mathbf{x_k}) = \min_{\pi_{k+1}} \sum_{\mathbf{x_{k+1}}} P(\mathbf{x_{k+1}}|\mathbf{x_k},\pi_{k+1})[d_{k+1}(\mathbf{x_k},\pi_{k+1}) + h_{k+1}(\mathbf{x_{k+1}})],$$

$$h_{n+1}(\mathbf{x_{n+1}}) = 0, \tag{2.42}$$

where
   $P(\mathbf{x_{k+1}}|\mathbf{x_k},\pi_{k+1})$ is the transition probability from state $\mathbf{x_k}$ to state $\mathbf{x_{k+1}}$
   $d_{k+1}(\mathbf{x_k},\pi_{k+1})$ is the transition time from state $\mathbf{x_k}$ to state $\mathbf{x_{k+1}}$ by taking π_k $(k = n, n-1,...,0)$

The transition probability is approximated by the probability of the steady-state of the Markov channel, that is, $p_{BG}/(p_{GB}+p_{BG})$ and $p_{GB}/(p_{GB}+p_{BG})$ for the channel state being good and bad, respectively. The transition time d_k can be either deterministic if the location indicators of state $\mathbf{x_k}$ and state $\mathbf{x_{k+1}}$ are the same, or stochastic if the location indicators are different with the input/output data transmission. For the former, we can compute the transition time by Equations 2.29 and 2.31. For the latter, we can approximate the transmission time by simulating the data transmission for a large number of times for Equations 2.33 and 2.35.

Similarly, we can have the backward value iteration for the minimum expected energy cost,

$$f_k(\mathbf{x_k}) = \min_{\pi_{k+1}} \sum_{\mathbf{x_{k+1}}} P(\mathbf{x_{k+1}}|\mathbf{x_k}, \pi_{k+1})[e_{k+1}(\mathbf{x_k}, \pi_{k+1}) + f_{k+1}(\mathbf{x_{k+1}})],$$
$$f_{n+1}(\mathbf{x_{n+1}}) = 0, \tag{2.43}$$

where

$f_k(\mathbf{x_k})$ is the minimum expected energy cost from state $\mathbf{x_k}$ at stage k to state $\mathbf{x_{n+2}}$ at stage $n + 2$

$f_0(\mathbf{x_0})$ is the minimum expected energy consumption to complete the entire application

We can also have the backward value iteration for the minimum expected aggregated cost,

$$J_k(\mathbf{x_k}) = \min_{\pi_{k+1}} \sum_{\mathbf{x_{k+1}}} P(\mathbf{x_{k+1}}|\mathbf{x_k}, \pi_{k+1})[e_{k+1}(\mathbf{x_k}, \pi_{k+1})$$
$$+ \lambda d_{k+1}(\mathbf{x_k}, \pi_{k+1}) + J_{k+1}(\mathbf{x_{k+1}})], \tag{2.44}$$
$$J_{n+1}(\mathbf{x_{n+1}}) = 0,$$

where

$J_k(\mathbf{x_k})$ is the minimum expected aggregated cost from state $\mathbf{x_k}$ at stage k to state $\mathbf{x_{n+2}}$ at stage $n + 2$

$J_0(\mathbf{x_0})$ is the minimum expected aggregated cost to complete the entire application

To this end, we can implement the procedure *ShortestPath* for time delay, energy cost, and aggregated cost using Equations 2.42 through 2.44, respectively, and then obtain the task scheduling policy for the collaborative task execution in Algorithm 2.2.

2.3.2.4 Performance Evaluation

For the performance evaluation, we compare collaborative task execution with two other execution strategies under the delay deadline constraint, that is, local execution and remote execution:

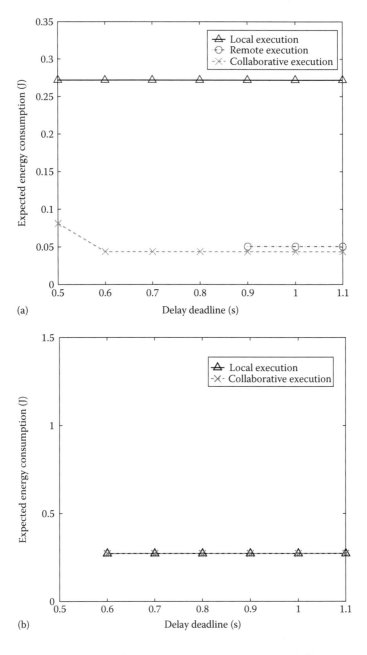

Figure 2.18   Comparison of energy consumption among different execution modes. (a) IID model ($\mathbb{E}(R) = 100$ kb/s), (b) IID model ($\mathbb{E}(R) = 10$ kb/s). (*Continued*)

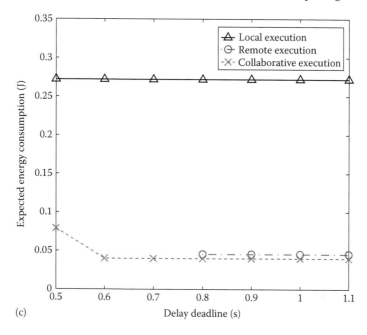

(c)

Figure 2.18 (*Continued*)   Comparison of energy consumption among different execution modes. (c) GE model ($p_{GG} = 0.995$, $p_{BB} = 0.96$, $R_G = 100$ kb/s and $R_B = 10$ kb/s).

- Local execution: All the tasks are executed locally on the mobile device.

- Remote execution: All the tasks are off-loaded to the cloud for execution.

We suppose that there are 10 tasks in the application. The workload is $\omega = \{30, 25, 16, 32, 15, 37, 44, 24, 40, 9\}$M cycles. The input and output data are $\alpha = \{22, 30, 6, 47, 30, 5, 47, 14, 49, 18\}$ kb and $\beta = \{30, 6, 47, 30, 5, 47, 14, 49, 18, 47\}$ kb, respectively.

We plot in Figure 2.18 the minimum expected energy consumption as a function of the application completion time constraint under the IID channel model and the Markovian channel model. The merits of the collaborative execution can be summarized as follows. First, the collaborative execution consumes less energy than the local execution and the remote execution. In Figure 2.18a and 2.18c, energy consumed by the mobile execution is about five times of that by the collaborative execution, under the IID model with high expected data rate and the GE model with the initial channel state to be good. In Figure 2.18b, however, the collaborative task execution becomes the local execution under the low expected data rate. This indicates that all the tasks are executed on the mobile device to avoid the data transmission for energy saving. Compared to the remote execution, the collaborative execution also results in less energy consumption, as shown in Figure 2.18a and 2.18c. Second, the collaborative execution is more flexible than the remote execution. In Figure 2.18b, the remote execution exceeds the delay deadline. In Figure 2.18a and 2.18c, the remote execution is only applicable if the delay deadline is no less than 0.8 and 0.9 s, respectively. This indicates that under the limited expected data rate or small delay deadline, we cannot use the remote execution. Therefore, the collaborative execution is more flexible and energy-efficient due to its fine granularity on task scheduling.

2.3.3 Application as a General Topology

In this section, we investigate the energy-efficient off-loading policy of collaborative execution for the application as a general topology.

2.3.3.1 Task Model and Problem Formulation

The tasks in general topology can be represented by a directed acyclic graph $G = (\mathcal{V}, \mathcal{A})$, where

$\mathcal{V}$ is the node set denoting the tasks, $\mathcal{V} = \{v_i\}$; we use the terms node and task interchangeably;

$\mathcal{A}$ is the arc set representing data dependencies among the tasks. An arc a_{ij} in $\mathcal{A}$ indicates the dependency between the adjacent task i and task j, such that task j cannot be started if its parent, task i, is not completed.

Suppose there are n tasks in the application. We add two dummy nodes into the graph, that is, v_0 and v_{n+1}, to denote the initialization and termination of the application. Figure 2.19 illustrates an example of graph representation of tasks in general topology.

We aim to find a schedule $\Pi = \{\pi_i\}$ that minimizes the energy consumption on the mobile device, while completing all the tasks within the delay deadline. The total energy consumption is the sum of the

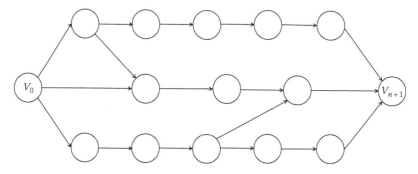

Figure 2.19   An example of a graph representation of tasks within an application.

weights of nodes and arcs in the graph G. The execution time of the application is when the last task in the graph (i.e., v_{n+1}) completes. This can be formulated as a delay constrained workflow scheduling problem. However, it is NP-complete [1]. Efficient algorithms are required for the workflow scheduling.

The workflow scheduling problem can also be formulated as an integer linear program. However, solving the integer linear program may not be efficient in practice due to its computation complexity. Inspired by the results of *one-climb* policy of the linear chain, we can decompose the general topology into a set of linear chains. This decomposition allows us to design a heuristic algorithm of the workflow scheduling.

2.3.3.2   Workflow Scheduling Algorithm

We leverage partial critical path analysis [2] and *one-climb* policy to design the algorithm for the workflow scheduling.

First, for the overall workflow, we adopt the partial critical path analysis. Specifically, the critical parent of a task is defined as its parent node that results in the maximum value of the earliest start time, under which the task can be started. By finding a set of critical parents, we can constitute a partial critical path PCP. We also add two tasks that have been scheduled into the PCP as the first and the last task of the PCP, respectively, such that we can identify the subdeadline of the *PCP*. By partial critical path analysis, the general topology can be decomposed into a set of paths, each of which is a linear chain case.

The earliest start time (*EST*) can be computed iteratively, that is,

$$EST(v_0) = 0$$
$$EST(v_i) = \max_{v_j \in \mathcal{P}(v_i)} \{EST(v_j) + MET(v_j) + CT(v_j, v_i, \pi_j, \pi_i)\}, \quad (2.45)$$

where
    $\mathcal{P}(v_i)$ is the set of parent tasks of v_i
    $MET(v_j)$ is the minimum execution time of task j
    $CT(v_j, v_i, \pi_j, \pi_i)$ is the communication time between task j and
    task i

Second, for each of the paths, we adopt the *one-climb* policy to schedule the execution of the tasks. The strategy is to minimize the energy consumption under the subdeadline, by determining the execution decision of each task on the *PCP*, that is, mobile execution or cloud execution. From the path set under the *one-climb* policy, we can choose the execution decision with the minimum energy consumption under the delay constraint.

The general framework of the workflow scheduling algorithm is illustrated in Algorithm 2.3. Given a graph $G = (\mathcal{V}, \mathcal{A})$, we can find the execution decision Π. The algorithm terminates when all of the tasks have been scheduled. By exploring the granularity of the tasks, we

---

Algorithm 2.3 General framework of the workflow scheduling algorithm

---

Require: $G(\mathcal{V}, \mathcal{A})$
Ensure: Schedule Π
  1: set tasks v_0 and v_{n+1} as scheduled
  2: **while** there exists an unscheduled task **do**
  3:     find partial critical path *PCP*
  4:     adopt *one-climb* policy to execute the tasks on the *PCP*
  5:     set the tasks on the *PCP* as scheduled
  6: **end while**

---

can have an energy-efficient collaborative task execution for mobile applications.

2.4   Policy in Task Delegation

In this section, we introduce the dispatching policy in task delegation. Particularly, we will present its application in TaaS. A transcoding task can be executed on the mobile device or off-loaded and scheduled by the dispatcher to one of the service engines in the cloud. We aim to minimize the energy consumption of transcoding on both the mobile device and service engines in the cloud.

2.4.1   Transcoding as a Service

Transcoding technology [43] has been an effective solution to adapt the videos based on the available bandwidth and device capability. Specifically, to be viewed smoothly by users, videos are transcoded into multiple rates, resolutions or formats for different devices. Such a transcoding operation, however, is computation-intensive for the resource-constrained mobile device. Therefore, it has become a research question on how to support video transcoding in both academia and industry.

With the emergence of cloud computing, TaaS [15,32] has been proposed for large scale computing in the field of multimedia. A generic cloud-based transcoding system is illustrated in Figure 2.20. Mobile devices, with different screen sizes and playback capabilities, can off-load the transcoding operation to the cloud for execution. In the cloud, there is a dispatcher at the front-end that receives off-loading requests from mobile devices, and a set of service engines and data storage in the back-end that stores the original video contents. If the video requested by users is cached in service engines, the cached video can be rendered immediately to users without transcoding; otherwise, video transcoding is required, either to be performed on the mobile device locally (i.e., mobile execution) or one of the service engines in the cloud (i.e., cloud execution). This requires us to design a task delegation policy on how to reduce the energy consumption on both the mobile device and the service engines in the cloud.

Previous works have presented design and mechanism for TaaS. Gao et al. in [13] proposed a partial transcoding scheme to minimize the long-term operational cost in renting VM instances from public

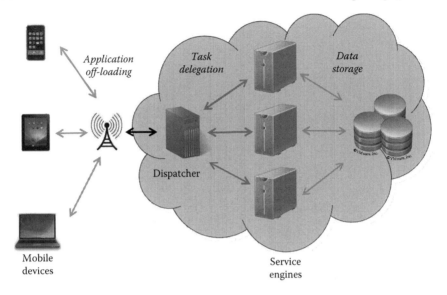

Figure 2.20 A generic cloud-based transcoding system. There is a dispatcher at the front-end that receives off-loading requests from mobile devices. There are also a set of service engines and data storage in the back-end.

cloud for user requests. Jin et al. in [23] proposed a PAINT (Partial In-Network Transcoding) scheme to reduce the transcoding cost and bandwidth cost in delivering video services over Information Centric Network (ICN). Jin et al. in [22] proposed a cloud clone migration scheme to minimize monetary cost in multi-screen cloud social TV systems, while enhancing the user experience for mobile users. In this chapter, we consider the energy consumption as the design metric for the purpose of green mobile cloud multimedia. Since we focus on the task delegation for video transcoding, the content placement and mobile video streaming in cloud media are not within the scope of this chapter; interested readers can refer to the works [17,19,24,58].

2.4.2   System Models and Problem Formulation

2.4.2.1   Arrival of Transcoding Tasks

We consider the arrival of transcoding tasks based on a discrete time slot model. The length of a time slot τ is assumed to be small such that for each time slot, there is at most one transcoding task arriving to the dispatcher. In addition, we denote p as the probability of one arrival to the dispatcher for each time slot and $1 - p$ if there are no arrivals.

Upon receiving the transcoding task, the dispatcher needs to schedule the task to one of the service engines for execution. We denote $u(t)$ as the decision variable. The decision variable is chosen from the set $\mathcal{U} = \{0, 1, 2, ..., N\}$, given by,

$$u(t) = \begin{cases} 0, & \text{if no arrival occurs} \\ i, & \text{if dispatched to service engine } i, \end{cases} \qquad (2.46)$$

where $i = 1, 2, ..., N$ and N is the number of service engines. Intuitively, if there are no arrivals, the dispatcher does not need to make the decision. Otherwise, the dispatcher decides the routing of the arriving task.

When the service engine receives the transcoding task from the dispatcher, the transcoding time for the task is associated with its constant CPU speed. The CPU speed s_i can be different for service engines. The service engine in faster CPU speed can have less completion time for transcoding. Without loss of generality, we assume that $s_1 \leq s_2 \leq ... \leq s_N$. We denote $A(t)$ as the transcoding time for the arrival at time slot t by a baseline server, which has a CPU speed S. The transcoding time $A(t)$ can be estimated by statistical learning and it is not the scope of this chapter. Suppose that the transcoding task is dispatched to the ith service engine. Then, the transcoding time at the ith service engine will be $A_i(t) = SA(t)/s_i$. Since the transcoding time of the same arrival (i.e., $A_i(t)$) can be different for service engines, the dispatcher needs to decide which service engine should process the transcoding task.

2.4.2.2   Queueing Dynamics of Service Engines

To measure the dynamics of service engines, we define queue length $\mathbf{Q}(t)$ as the remaining transcoding time of tasks in service engines at time slot t, that is, $\mathbf{Q}(t) = \{Q_1(t), Q_2(t), ..., Q_N(t)\}$. The queue length $\mathbf{Q}(t)$ reflects the queueing delay on service engines. Specifically, the queue length of the ith service engine evolves according to

$$Q_i(t+1) = \max[Q_i(t) - \tau, 0] + A_i(t)\mathbf{1}_{\{u(t)=i\}}, \qquad (2.47)$$

where $\mathbf{1}_{\{u(t)=i\}}$ is an indicator function that is 1 if $u(t) = i$ and 0 otherwise. If the arrival is dispatched to the ith service engine (i.e., $u(t) = i$), its queue length will be increased by $A_i(t)$; otherwise, no arrival occurs to the ith service engine. Based on Equation 2.47, the dispatcher can observe the queue length of service engines for each time slot.

To consider the long-term performance, we define the time average queue length as

$$\overline{Q} = \lim_{T \to \infty} \sup \frac{1}{T} \sum_{t=0}^{T-1} \sum_{i=1}^{N} \mathbb{E}\{Q_i(t)\}, \tag{2.48}$$

where the expectation is taken over the randomness of $A(t)$.

2.4.2.3   Energy Consumption on Mobile Devices

We can treat the application of video transcoding on mobile devices as an example of the one node case, and leverage the energy consumption model for the video transcoding. For the mobile execution, the energy consumption on the mobile device is modeled the same as that in the previous section. For the cloud execution, the energy consumption on the mobile device consists of the energy consumed by transmitting the input data and receiving the output data. First, the input data L is transmitted to the cloud by adapting the transmission rate r_t, and the resulted energy consumption is a monomial function of the data transmitted (described in the one node case). Second, the output data L' is received at a constant rate r' and power P'. Thus, the total energy consumption of the cloud execution is $\mathbb{E}\{\mathcal{E}_{tran}(L, T_s, \phi)\} + \mathcal{E}_{recv}(L')$, where $\mathcal{E}_{tran}$ and $\mathcal{E}_{recv}$ are the energy consumed by transmitting the input data and receiving the output data, respectively. Moreover, T_s is the transmission delay, given by $T_s = T_d - (L'/r') - T_0$, where T_0 is the queueing delay, depending on the off-loading policy for which service engine is chosen to perform the transcoding task. Since the receiving energy consumption $\mathcal{E}_{recv}$ is a constant, we only need to find the data transmission schedule $\phi = \{r_1, r_2, ... r_{T_s}\}$ under the transmission delay T_s to minimize the transmission energy consumption $\mathcal{E}_{tran}$.

2.4.2.4   Energy Consumption on Service Engines

The energy consumption on service engines can be modeled as a function of the CPU speed for processing transcoding tasks. If the dispatcher dispatches the transcoding task to the ith service engine at time slot t, the energy consumption on the ith service engine is the product of the transcoding time and the power $A_i(t)\kappa s_i^\gamma$, where the power κs_i^γ is a convex function of CPU speed s_i [9,41]. Normally, γ is set to be 3 [9]. In addition, without loss of generality, we set the constant parameter $\kappa = 1$. If the service engine has no transcoding tasks to process, the service engine can be set to sleep mode, resulting in very small energy consumption that can be negligible. Thus, the energy consumption for completing the task is

$$E_i(t) = A_i(t)\kappa s_i^\gamma \mathbf{1}_{\{u(t)=i\}}, \tag{2.49}$$

where $\mathbf{1}_{\{u(t)=i\}}$ is the indicator function that denotes 1 if $u(t) = i$ and 0 otherwise. For the long-term performance, we consider the time average energy consumption, given by

$$\overline{E} = \lim_{T \to \infty} \sup \frac{1}{T} \sum_{t=0}^{T-1} \sum_{i=1}^{N} \mathbb{E}\{E_i(t)\}, \tag{2.50}$$

where the expectation is taken over the randomness of $A(t)$.

We regard each service engine as a physical machine. Particularly, we only consider the computation energy consumption in the service engine. The justifications can be given as follows. First, computation energy is a dominant term in the distributed servers [4,6], and the other components of energy consumption (such as memory and network) can be ignored. Second, the computation energy is larger than the energy caused by the transition overhead from the sleep mode to the running mode. Specifically, for the computation-intensive transcoding tasks, the computation energy and time are the first-order component for the total energy and time, while the energy and time due to the transition overhead are the second-order component.

For the virtual machine, its energy consumption model can be more complex. One can adjust the energy consumption model and then adopt the mathematical framework for the energy-efficient task execution.

2.4.2.5 Problem Formulation

For the mobile device, it has two options for the transcoding operation, that is, mobile execution and cloud execution. To determine which one is more energy-efficient, we can formulate delay-constrained optimization problems for the mobile execution and the cloud execution, respectively. For the mobile execution, its formulation has been presented in Section 2.3.1. For the cloud execution, its formulation is defined as follows,

$$\mathcal{E}_c^* = \min_{\phi \in \Phi} \mathbb{E}\{\mathcal{E}_{tran}(L, T_s, \phi)\} + \mathcal{E}_{recv}(L'), \tag{2.51}$$

where we have considered the receiving time for the transcoding operation.

For service engines in the cloud, we determine which service engine should perform the transcoding task upon its arrival. We can formulate a stability-constrained optimization problem to minimize the time average energy consumption, while satisfying the queue stability,

$$\min_{\{u(t)\}} \quad \overline{E}, \tag{2.52}$$

$$\text{s.t.} \quad \overline{Q} < \infty, \tag{2.53}$$

$$u(t) \in \mathcal{U}, \tag{2.54}$$

where $u(t)$ is the decision variable, Equation 2.53 denotes the queue stability constraint and Equation 2.54 denotes the feasibility constraint.

> To guarantee the delay of transcoding tasks, we require all the queues to be stable. According to Little's theorem, the average queue length is proportional to the average delay. The constraint of Equation 2.53 indicates that time average queue length should not go to infinity, and thus the time average delay is finite.

2.4.3 Off-Loading Policy for Mobile Devices

We can have off-loading policy for mobile devices based on the results in Section 2.3.1. The minimum energy consumption of the mobile device by the mobile execution is $\mathcal{E}_m^* = ML^3/T_d^2$, and the minimum energy consumption of the mobile device by the cloud execution is

$$\mathcal{E}_c^* = \frac{C_0(n)L^n}{(T_d - L'/r' - T_0)^{n-1}} + \frac{P'L'}{r'}, \tag{2.55}$$

where the first term on the right-hand side of the equation refers to the energy consumption of transmitting the input data and the second term refers to the energy consumption of receiving the output data. In addition, $C_0(n)$ is a function of monomial order n for the cloud execution.

Example 2.12 *We consider the application of transcoding high-definition videos of FLV files with 1920 ×1080 resolution size into mp4 files with 320 × 240 resolution size. We collect both the input and output data, and model the output data size L' as a linear function of input data size L. Figure 2.21 shows the fitting model $L' = 0.0175L + 0.3093$.*

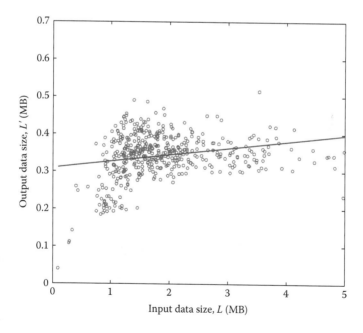

Figure 2.21   Curve fitting of output data size given input data size for video transcoding. The output data size L' can be modeled as a linear function of the input data size L.

Given input data size L and delay deadline T_d, we compare the optimal energy consumption by the mobile execution and the cloud execution, that is, $\mathcal{E}_m^*$ and $\mathcal{E}_c^*$, and plot the operational regions for the two execution modes, respectively, in Figure 2.22. Particularly, the operational regions are given under $n = 2$. First, Figure 2.22a and b shows the cases under the same queueing delay $T_0 = 0.5\,s$, but different receiving rate r'. If the receiving rate is larger (i.e., $r' = 8\,Mb/s$ in Figure 2.22b), the region of the cloud execution is larger; hence, it is more likely to accomplish the transcoding task on the cloud clone. This is because, with a larger receiving rate, the transmission time is longer under a total application completion deadline, such that the mobile device can slow down the data rate for transmission, resulting in less energy. Second, Figure 2.22c and d shows the cases under the same receiving rate $r' = 4\,Mb/s$, but different queueing delay. If the queueing delay is shorter (i.e., $T_0 = 0.1\,s$), the region of the cloud execution becomes larger. This is because, with a shorter queueing delay, the transmission time is longer under a total application completion deadline such that the mobile device can consume less energy in slowing down the data rate for transmission. In addition, it seems that the operational regions are separated by a line under large receiving rate and small queueing delay, as illustrated in Figure 2.22b and d, respectively. This is because, given the short queueing delay and the receiving time, the transmission time will be close to the total application completion deadline.

2.4.4 Dispatching Algorithm for Service Engines

We consider how to dispatch transcoding tasks to a set of available service engines to save energy consumption. We leverage the Lyapunov optimization to solve the optimization problem Equation 2.52 and design an online dispatching algorithm called REQUEST, which can Reduce the Energy consumption on service engines while achieving the QUEue STability.

For the ease of analysis, we first assume that transcoding time of an arriving task by the baseline server $A(t)$ is independent and identically distributed (i.i.d.) for every time slot. Then, $A_i(t)$ is also i.i.d. for the ith service engine.[*]

---

[*] The obtained results can also be extended to the non-i.i.d. model and the i.i.d. assumption is not crucial to the analysis.

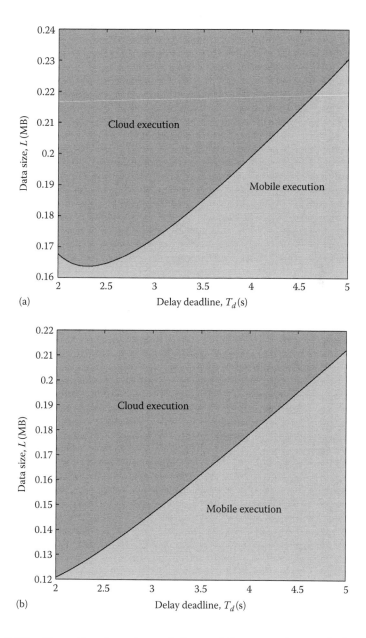

Figure 2.22   Operational region of the mobile execution and the cloud execution for mobile devices, where $n = 2$. (a) $T_0 = 0.5$ s. $r' = 4$ Mb/s, (b) $T_0 = 0.5$ s. $r' = 8$ Mb/s.   (*Continued*)

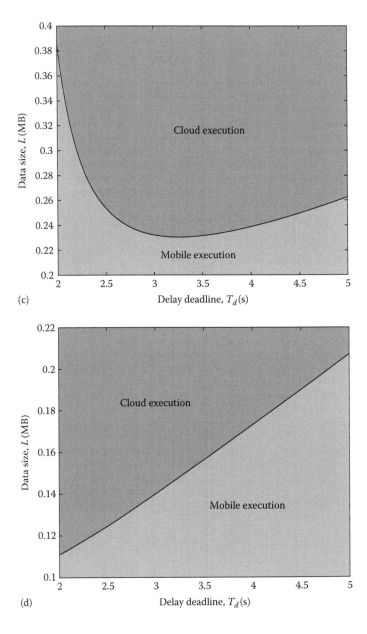

Figure 2.22 (*Continued*)   Operational region of the mobile execution and the cloud execution for mobile devices, where $n = 2$. (c) $T_0 = 1$ s. $r' = 4$ Mb/s, (d) $T_0 = 0.1$ s. $r' = 4$ Mb/s.

Under the framework of Lyapunov optimization, we define the quadratic Lyapunov function

$$L(\mathbf{Q}(t)) = \frac{1}{2} \sum_{i=1}^{N} Q_i(t)^2. \tag{2.56}$$

To capture the expected changes of the Lyapunov function from one slot to another, we define the conditional one-slot Lyapunov drift $\Delta(\mathbf{Q}(t)) = \mathbb{E}\{L(\mathbf{Q}(t+1)) - L(\mathbf{Q}(t))|\mathbf{Q}(t)\}$, where

$$\begin{aligned}
L(\mathbf{Q}(t+1)) - L(\mathbf{Q}(t)) &= \frac{1}{2} \sum_{i=1}^{N} Q_i(t+1)^2 - \frac{1}{2} \sum_{i=1}^{N} Q_i(t)^2 \\
&= \frac{1}{2} \sum_{i=1}^{N} \{\max[Q_i(t) - \tau, 0] + A_i(t)\mathbf{1}_{\{u(t)=i\}}\}^2 \\
&\quad - \frac{1}{2} \sum_{i=1}^{N} Q_i(t)^2.
\end{aligned} \tag{2.57}$$

Suppose all $A_i(t)$ are upper bounded by A_{max} (i.e., $A_i(t) \le A_{max}$). Using the fact $(\max[x-y,0] + z)^2 \le x^2 + y^2 + z^2 + 2x(z-y)$ for $\forall x, y, z \ge 0$, we have

$$\begin{aligned}
L(\mathbf{Q}(t+1)) - L(\mathbf{Q}(t)) &\le \frac{1}{2}[A_{max}^2 + N\tau^2] \\
&\quad + \sum_{i=1}^{N} Q_i(t)A_i(t)\mathbf{1}_{\{u(t)=i\}} - \tau \sum_{i=1}^{N} Q_i(t).
\end{aligned} \tag{2.58}$$

Therefore, we have the bound for the Lyapunov drift:

$$\Delta(\mathbf{Q}(t)) \le B - \mathbb{E}\left\{ \sum_{i=1}^{N} \tau Q_i(t)|\mathbf{Q}(t) \right\} + \mathbb{E}\left\{ \sum_{i=1}^{N} Q_i(t)A_i(t)\mathbf{1}_{\{u(t)=i\}}|\mathbf{Q}(t) \right\}, \tag{2.59}$$

where B is a finite constant satisfying $B = \frac{1}{2}(A_{max}^2 + N\tau^2)$.

As stated in [37], the quadratic form of Lyapunov function indicates that if $L(\mathbf{Q}(t))$ is small, then all the queue lengths are small; if $L(\mathbf{Q}(t))$ is large, then at least one queue length is large.

The conditional one-slot Lyapunov drift expresses the expected changes of the Lyapunov function from one slot to the next. It is shown by [37] that the minimization of the right hand side of Equation 2.59 will guarantee the queue stability.

To achieve the minimum energy consumption on service engines, we consider the *drift-plus-penalty* function for the dispatching algorithm, which is a weighted sum of the drift and the penalty,

$$F(\mathbf{Q}(t)) = \Delta(\mathbf{Q}(t)) + V\mathbb{E}\{E(t)|\mathbf{Q}(t)\}, \tag{2.60}$$

where $V \geq 0$. In particular, the penalty is $E(t) = \sum_{i=1}^{N} E_i(t)$. Plugging Equation 2.59 into Equation 2.60, we can have the bound of the *drift-plus-penalty* function:

$$F(\mathbf{Q}(t)) \leq B - \mathbb{E}\left\{\sum_{i=1}^{N} \tau Q_i(t)|\mathbf{Q}(t)\right\} + \mathbb{E}\left\{\sum_{i=1}^{N} Q_i(t)A_i(t)\mathbf{1}_{\{u(t)=i\}}|\mathbf{Q}(t)\right\}$$
$$+ V\mathbb{E}\left\{\sum_{i=1}^{N} A_i(t)\kappa s_i^{\gamma}\mathbf{1}_{\{u(t)=i\}}|\mathbf{Q}(t)\right\}. \tag{2.61}$$

We minimize the righthand side of Equation 2.61 and design the dispatching algorithm in Algorithm 2.4. If $V = 0$, only drift function is considered and we choose the service engine with the minimum $Q_i(t)A_i(t) = Q_i(t)(A(t)S/s_i)$, that is, $Q_i(t)/s_i$ determines the policy, which results in shorter queue and faster server basis. If $V \neq 0$, the *drift-plus-penalty* function is considered and we choose the service engine with the minimum $A_i(t)(Q_i(t) + V\kappa s_i^{\gamma})$, which demands the trade-off between the queue length and the energy consumption.

We can characterize the trade-off between the time average energy consumption and the time average queue length in Theorem 2.5. The derivation of Theorem 2.5 can be found in [54].

Algorithm 2.4 REQUEST Algorithm

Require: Q(t)

Ensure: $u(t)$

1: At the beginning of each time slot t, the dispatcher observes the queue length of service engines **Q**(t).

2: The dispatcher finds $u(t)$ to minimize the following function

$$\sum_{i=1}^{N} Q_i(t)A_i(t)\mathbf{1}_{\{u(t)=i\}} + V\sum_{i=1}^{N} A_i(t)\kappa s_i^{\gamma}\mathbf{1}_{\{u(t)=i\}} \qquad (2.62)$$

i.e.,

$$u(t) = \arg\min_i\{A_i(t)(Q_i(t) + V\kappa s_i^{\gamma})\}. \qquad (2.63)$$

3: Service engines update the queue length **Q**(t) according to Equation 2.47.

Theorem 2.5 *Suppose that the arrival of transcoding tasks is within capacity region, where the capacity region requires*

$$\mathbb{E}\{a_i(t)\} < \tau, \ \forall i, \qquad (2.64)$$

with $a_i(t) = A_i(t)\mathbf{1}_{\{u(t)=i\}}$. Then, given that $\mathbb{E}\{L(\mathbf{Q}(0))\} < \infty$ and control variable $V > 0$, the online algorithm can stabilize the system. The resulted time average energy consumption and queue length will satisfy the following inequalities:

$$\overline{E} \leq E^* + \frac{B}{V}, \qquad (2.65)$$

$$\overline{Q} \leq \frac{B + VE^*}{\varepsilon}, \qquad (2.66)$$

where

ε is a constant

E^* is a lower bound on the time average energy consumption

Capacity region ensures that the expected transcoding time by a service engine for one time slot should be less than the length of one time slot. Indeed, if we multiply s_i on both sides of Equation 2.64, we will find that the incoming expected workload to the ith service engine should not exceed the capacity of the ith service engine.

Theorem 2.5 shows that we can achieve the energy-delay trade-off by controlling the variable V. If we choose a large value for the control variable V, the time average energy consumption $\overline{E}$ can be arbitrarily close to the optimal one E^*. However, this will render the time average queue length to increase linearly with V, such that a long delay will incur. The trade-off between the time average consumption and time average queue length is critical for the cloud operator. In addition, we can extend these results for the Markovian transcoding time $A(t)$ and the $[O(1/V), O(V)]$ energy-delay trade-off is still preserved, that is,

$$\overline{E} \leq E^* + O\left(\frac{1}{V}\right), \tag{2.67}$$

$$\overline{Q} \leq O(V). \tag{2.68}$$

This can be proved by using multi-slot drift analysis [37].

Example 2.13 *We illustrate an example for Theorem 2.5.*

First, we measure the elapsed time of video transcoding by FFmpeg. Particularly, we transcode a set of flv files into mp4 files in six commonly used resolution cases, that is, 320×240, 427×240, 480×360, 640×360, 640×480, and 854×480. The original high-definition videos are in 1920×1080, with equal duration time (5 s) but different file size (ranging from 0.1 to 5 MB, with the mean 1.87 MB). Here, we only consider the case of transcoding videos into different resolution sizes and formats. The optimization framework is still valid for the case of adapting the bit rate of videos.

Second, we model the transcoding time as a function of the file size of a video. Particularly, we define $A = LX$, where L is the file size and X is a random variable that denotes transcoding time for a unit of file size. We find that X can be modeled by a gamma distribution, where its CDF fitting are shown in Figure 2.23 (with the shape parameter k and scale parameter θ). Note that the dispatching algorithm is not restricted to this model, but can be extended for other statistical models of transcoding time.

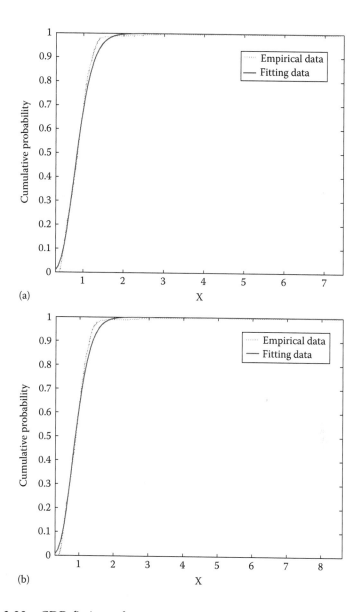

Figure 2.23 CDF fitting of gamma distribution for transcoding time in six resolution cases. (a) 320 × 240: $k = 7.8270$, $\theta = 0.1129$; (b) 427 × 240: $k = 7.4772$, $\theta = 0.1244$.   (*Continued*)

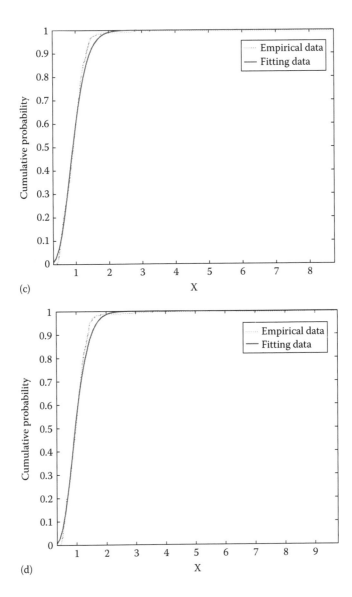

Figure 2.23 (*Continued*)   CDF fitting of gamma distribution for transcoding time in six resolution cases. (c) 480 × 360: $k = 7.2159$, $\theta = 0.1328$; (d) 640 × 360: $k = 7.2076$, $\theta = 0.1392$.   (*Continued*)

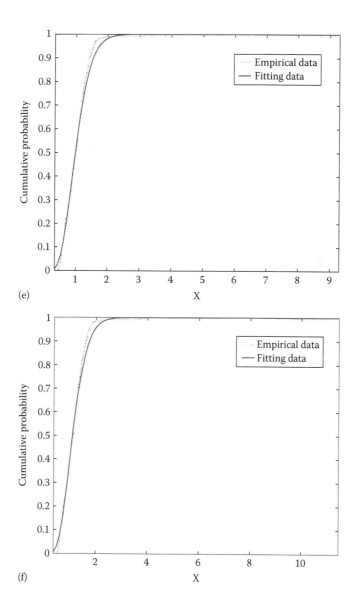

Figure 2.23 (*Continued*)   CDF fitting of gamma distribution for transcoding time in six resolution cases. (e) 640 × 480: $k = 7.2240$, $\theta = 0.1421$; (f) 854 × 480: $k = 6.7888$, $\theta = 0.1683$.

Then, we can have the simulation as follows. We set the length of the time slot $\tau = 0.5$ s. The CPU speed of a baseline server is $S = 3.2$ GHz (Intel(R) Xeon(R) CPU E5-1650 0), with 16 GB memory. Suppose that there are 10 service engines, and their CPU speeds range from 2.0 to 2.9 GHz with the increment of 0.1 GHz. We set $\kappa = 1$ and $\gamma = 3$ for the energy model. Each service engine has an empty queue at the first time slot. Then, we plot the trade-off between energy consumption and queue length in Figure 2.24. For each specific resolution case, with the increase of V, the time average energy consumption decreases and converges to the optimal value, while the time average queue length grows linearly. These results are consistent with Theorem 2.5.

2.4.5 Performance Evaluation

We compare the performance of REQUEST algorithm with two alternative algorithms, that is, Round Robin and Random Rate. Specifically, the Round Robin and Random Rate algorithms are illustrated as follows:

- Round Robin: Transcoding tasks are scheduled in circular order among the N service engines.

- Random Rate: Transcoding tasks are routed to the ith service engine with the probability $s_i / \sum_{i=1}^{N} s_i$, where s_i is the CPU speed of the ith service engine.

Round Robin and Random Rate algorithms work in a static manner to balance the load among service engines. However, they are unaware of the arrival transcoding tasks. We will show that they could not achieve small energy consumption.

We first investigate the trade-off between time average energy consumption and time average queue length for these three dispatching algorithms. Figure 2.25 shows that Round Robin and Random Rate have a fixed time average energy consumption and time average queue length, while the REQUEST algorithm is more adaptive to balance the trade-off between time average energy consumption and time average queue length.

We then compare these three algorithms and plot the time average energy consumption, the time average queue length, and the file size in each time slot that reflects the traffic under $p = 0.8$ in

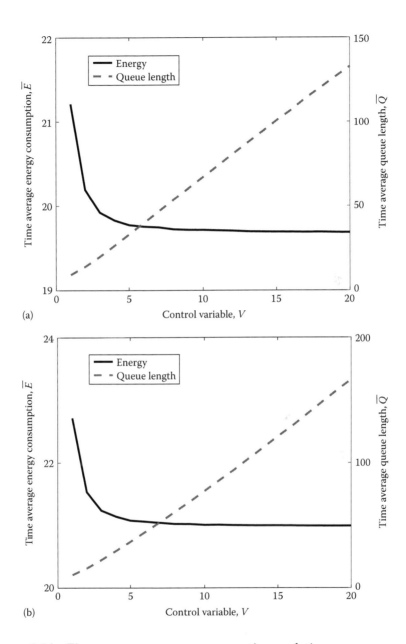

Figure 2.24 Time average energy consumption and time average queue length under different V values for various resolution cases. $V = \{1 : 1 : 20\}$. $T = 100{,}000$. $p = 0.8$. (a) 320×240, (b) 427×240.   (*Continued*)

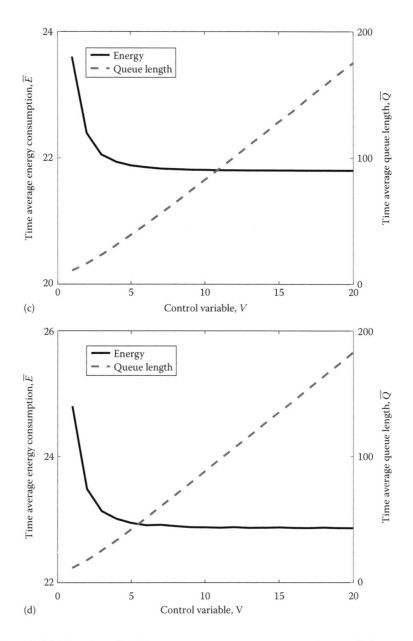

Figure 2.24 (*Continued*)   Time average energy consumption and time average queue length under different *V* values for various resolution cases. $V = \{1 : 1 : 20\}$. $T = 100,000$. $p = 0.8$. (c) 480×360, (d) 640×360.

(*Continued*)

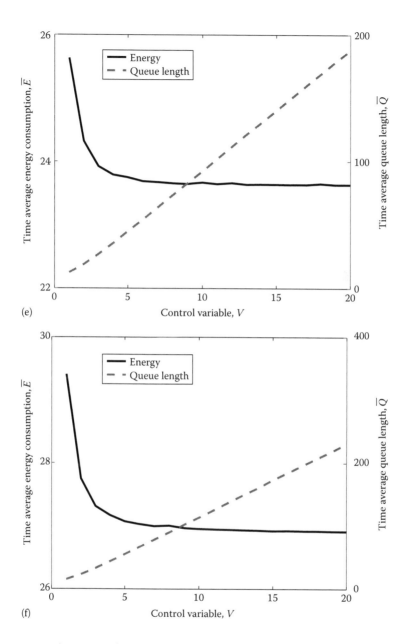

Figure 2.24 (*Continued*)   Time average energy consumption and time average queue length under different V values for various resolution cases. $V = \{1 : 1 : 20\}$. $T = 100,000$. $p = 0.8$. (e) 640×480, (f) 854×480.

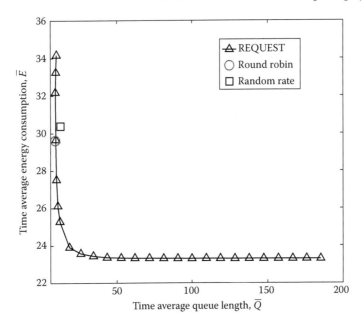

Figure 2.25 Trade-off between time average energy consumption and time average queue length for dispatching algorithms. $T = 100000$. $V = \{0, 0.07, 0.14, 0.3, 0.5, 0.75, 1 : 1 : 20\}$. $p = 0.8$.

Figure 2.26 from top to bottom, respectively. We set $V = 0, 1, 5$ in the REQUEST algorithm, respectively, for the comparison. Under $V = 0$, the REQUEST algorithm has close time average queue length to Round Robin and Random Rate algorithms, but with the highest time average energy consumption. Under $V = 1$ and $V = 5$, the REQUEST algorithm can have small time average energy consumption, which is about 17% and 27% smaller than Round Robin and Random Rate algorithms, respectively. Specifically, under $V = 5$, the REQUEST algorithm can achieve the smallest time average energy consumption, but its time average queue length is the largest. Under $V = 1$, the REQUEST algorithm achieves a slightly larger time average queue length than Round Robin and Random Rate algorithms, but it can still maintain the time average queue length to be small. Therefore, the cloud operator can tune the control variable V of the REQUEST algorithm to achieve the minimum energy consumption on service engines.

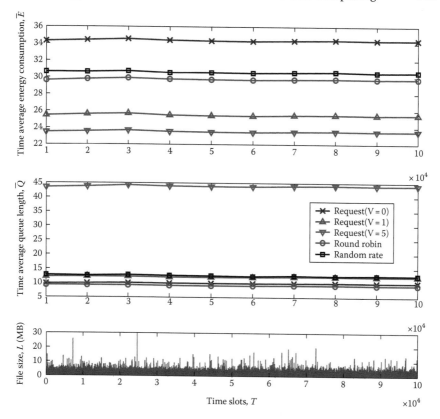

Figure 2.26 Performance comparison under the arrival probability $p = 0.8$.

2.5 Summary

This chapter presented the off-loading policy of task execution for computation-intensive applications in mobile cloud computing.

We first designed the off-loading policy based on the task granularity of the mobile application, including one node, linear chain, and general topology. For the application as a node, we obtained the optimal clock frequency configuration of the mobile execution and the optimal data transmission scheduling of the cloud execution, respectively. We then obtained the optimal operational region for the application execution, which is characterized by a threshold, depending on the wireless transmission model and the ratio of energy coefficients for mobile execution and cloud execution. For the application as a linear chain, we derived a *one-climb* policy by characterizing the optimal solution and proposed an enumeration algorithm for the collaborative task execution in polynomial time. Further, we adopted LARAC algorithm to solve the optimization problem approximately. For the application as a general topology, we leveraged the property of *one-climb* policy and partial critical path analysis to design the workflow scheduling algorithm for collaborative task execution.

We then designed the dispatching policy of a typical application in mobile cloud computing, that is, video transcoding, where we minimized the energy consumption of transcoding on the mobile devices and service engines in the cloud while achieving the queue stability. For the mobile device, we considered the queueing delay for the off-loading request and obtained the operational region of the optimal execution for mobile devices. For service engines in the cloud, we formulated the off-loading policy as a stability-constrained optimization problem. We leveraged Lyapunov optimization framework and proposed an online algorithm to dispatch transcoding tasks to service engines, which can reduce energy consumption while achieving the queue stability. The proposed policy is not limited to video transcoding, but could be also extended for other applications (e.g., image retrieval, virus scanning).

Glossary

Application Completion Probability (ACP): The probability that the application gets completed.

Dynamic Voltage Scaling (DVS): A power management in which the voltage can scale up and down.

Gilbert–Elliott channel model: A channel model where there are two states, that is, good and bad.

Lagrangian Aggregated Cost (LARAC): Lagrange relaxation based aggregated cost.

Markov Decision Process (MDP): A mathematical framework of modeling decision making.

One-climb policy: A policy for tasks in linear chain in mobile cloud computing.

Partial Critical Path (PCP): A method to find critical tasks in the graph.

REQUEST: An online algorithm that can reduce the energy consumption on service engines while achieving the queue stability.

Transcoding as a Service (TaaS): A cloud-based service model for video transcoding.

References

1. S. Abrishami, M. Naghibzadeh, and D. H. J. Epema. Deadline-constrained workflow scheduling algorithms for infrastructure as a service clouds. *Future Generation Computer Systems*, 29(1): 158–169, 2013.

2. S. Abrishami and M. Naghibzadeh. Deadline-constrained workflow scheduling in software as a service cloud. *Scientia Iranica*, 19(3):680–689, 2012.

3. P. Balakrishnan and C.-K. Tham. Energy-efficient mapping and scheduling of task interaction graphs for code offloading in mobile cloud computing. In *Proceedings of the IEEE/ACM Sixth International Conference on Utility and Cloud Computing*, pp. 34–41, 2013.

4. A. Berl, E. Gelenbe, M. Di Girolamo, G. Giuliani, H. De Meer, M. Q. Dang, and K. Pentikousis. Energy-efficient cloud computing. *The Computer Journal*, 53(7):1045–1051, 2010.

5. E. Biglieri, J. Proakis, and S. Shamai (Shitz). Fading channels: Information-theoretic and communications aspects. *IEEE Transactions on Information Theory*, 44(6):2619–2692, 1998.

6. P. Bohrer, E. N. Elnozahy, T. Keller, M. Kistler, C. Lefurgy, C. McDowell, and R. Rajamony. The case for power management in web servers. In *Power Aware Computing*, pp. 261–289. Springer, New York, 2002.

7. T. Burd and R. Broderson. Processor design for portable systems. *Journal of VLSI Singapore Process*, 13:203–222, 1996.

8. S. Chen, Y. Wang, and M. Pedram. A semi-markovian decision process based control method for offloading tasks from mobile devices to the cloud. In *IEEE Globecom*, Atlanta, GA, pp. 2885–2890, 2013.

9. Y. Chen, A. Das, W. Qin, A. Sivasubramaniam, Q. Wang, and N. Gautam. Managing server energy and operational costs in hosting centers. In *Proceedings of the ACM SIGMETRICS International Conference on Measurement and Modeling of Computer Systems*, New York, pp. 303–314. ACM, 2005.

10. B. G. Chun, S. h. Ihm, P. Maniatis, M. Naik, and A. Patti. Clonecloud: Elastic execution between mobile device and cloud. In *Proceedings of the Sixth European Conference on Computer Systems*, New York, pp. 301–314, ACM, 2011.

11. E. Cuervo, A. Balasubramanian, D. Cho, A. Wolman, S. Saroiu, R. Chandra, and P. Bahl. MAUI: Making smartphones last longer with code offload. In *International Conference on Mobile Systems, Applications, and Services*, San Francisco, CA, pp. 49–62, 2010.

12. H. T. Dinh, C. Lee, D. Niyato, and P. Wang. A survey of mobile cloud computing: Architecture, applications, and approaches. *Wireless Communications and Mobile Computing*, 13(18):1587–1611, 2013.

13. G. Gao, W. Zhang, Y. Wen, Z. Wang, W. Zhu, and Y. P. Tan. Cost-optimal video transcoding in media cloud: Insights from user viewing pattern. In *International Conference on Multimedia and Expo*, Chengdu, China, pp. 1–6. IEEE, 2014.

14. W. Gao, L.-Y. Duan, J. Sun, J. Yuan, Y. Wen, Y.-P. Tan, J. Cai, and A. C. Kot. Mobile media communication, processing, and analysis: A review of recent advances. In *IEEE International Symposium on Circuits and Systems*, Beijing, China, pp. 869–872, 2013.

15. A. Garcia, H. Kalva, and B. Furht. A study of transcoding on cloud environments for video content delivery. In *Proceedings of the Multimedia Workshop on Mobile Cloud Media Computing*, Firenze, Italy, pp. 13–18. ACM, 2010.

16. I. Giurgiu, O. Riva, D. Juric, I. Krivulev, and G. Alonso. Calling the cloud: Enabling mobile phones as interfaces to cloud

applications. In *Proceedings of the 10th ACM/IFIP/USENIX International Conference on Middleware*, Urbana, Champaign, IL, pp. 83–102, 2009.

17. J. He, Z. Xue, D. Wu, and Y. Wen. CBM: Online strategies on cost-aware buffer management for mobile video streaming. *IEEE Transactions on Multimedia*, 16(1):242–252, 2013.

18. G. Q. Hu, W. P. Tay, and Y. G. Wen. Cloud robotics: Architecture, challenges and applications. *IEEE Network Special Issue on Machine and Robotic Networking*, 26(3):21–28, 2012.

19. H. Hu, Y. Wen, T.-S. Chua, Z. Wang, J. Huang, W. Zhu, and D. Wu. Community based effective social video contents placement in cloud centric CDN network. In *IEEE International Conference on Multimedia and Expo*, Chengdu, China, pp. 1–6. IEEE, 2014.

20. D. Huang, P. Wang, and D. Niyato. A dynamic offloading algorithm for mobile computing. *IEEE Transactions on Wireless Communications*, 11(6):1991–1995, 2012.

21. G. Huerta-Canepa and D. Lee. A virtual cloud computing provider for mobile devices. In *Proceedings of the First ACM Workshop on Mobile Cloud Computing & Services: Social Networks and Beyond*, San Francisco, CA, p. 6, 2010.

22. Y. Jin, Y. Wen, H. Hu, and M. Montpetit. Reducing operational costs in cloud social TV: An opportunity for cloud cloning. *IEEE Transactions on Multimedia*, 16(6):1739–1751, 2014.

23. Y. Jin and Y. Wen. PAINT: Partial in-network transcoding for adaptive streaming in information centric network. In *Proceedings of IEEE/ACM International Symposium of Quality of Service*, Hongkong, China, pp. 208–217, 2014.

24. Y. Jin, Y. Wen, K. Guan, D. Kilper, and H. Xie. Toward monetary cost effective content placement in cloud centric media network. In *IEEE International Conference on Multimedia and Expo*, San Jose, CA, pp. 1–6. IEEE, 2013.

25. L. A. Johnston and V. Krishnamurthy. Opportunistic file transfer over a fading channel: A POMDP search theory formulation with optimal threshold policies. *IEEE Transactions on Wireless Communications*, 5(2):394–405, 2006.

26. A. Juttner, B. Szviatovski, I. Mécs, and Z. Rajkó. Lagrange relaxation based method for the QoS routing problem. In *Proceedings of IEEE International Conference on Computer Communications*, Alaska, Vol. 2, pp. 859–868, 2001.

27. R. Kemp, N. Palmer, T. Kielmann, F. Seinstra, N. Drost, J. Maassen, and H. Bal. eyedentify: Multimedia cyber foraging from a smartphone. In *Proceedings of the 11th IEEE International Symposium on Multimedia*, San Diego, CA, pp. 392–399, 2009.

28. K. Kumar and Y. H. Lu. Cloud computing for mobile users: Can offloading computation save energy? *IEEE Computer*, 43(4):51–56, 2010.

29. K. Kumar, Y. Nimmagadda, and Y.-H. Lu. Energy conservation for image retrieval on mobile systems. *ACM Transactions on Embedded Computing Systems*, 11(3):66, 2012.

30. Y.-W. Kwon and E. Tilevich. Energy-efficient and fault-tolerant distributed mobile execution. In *Proceedings of the 32nd International Conference on Distributed Computing Systems*, Macau, China, pp. 586–595. IEEE, 2012.

31. J. Lee and N. Jindal. Energy-efficient scheduling of delay constrained traffic over fading channels. *IEEE Transactions on Wireless Communications*, 8(4):1866–1875, 2009.

32. Z. Li, Y. Huang, G. Liu, F. Wang, Z.-L. Zhang, and Y. Dai. Cloud transcoder: Bridging the format and resolution gap between internet videos and mobile devices. In *Proceedings of the 22nd International Workshop on Network and Operating System Support for Digital Audio and Video*, Toronto, ON, Canada, pp. 33–38. ACM, 2012.

33. X. Lin, Y. Wang, and M. Pedram. An optimal control policy in a mobile cloud computing system based on stochastic data. In *Proceedings of the Second International Conference on Cloud Networking*, San Francisco, CA, pp. 117–122. IEEE, 2013.

34. J. R. Lorch and A. J. Smith. Improving dynamic voltage scaling algorithms with PACE. In *Proceedings of ACM SIGMETRICS 2001*, Cambridge, MA, pp. 50–61, 2001.

35. A. P. Miettinen and J. K. Nurminen. Energy efficiency of mobile clients in cloud computing. In *Proceedings of the Second USENIX Conference on Hot Topics in Cloud Computing*, Boston, MA, pp. 4–11, 2010.

36. M. J. Neely, E. Modiano, and C. E. Rohrs. Dynamic power allocation and routing for time varying wireless networks. In *Proceedings of IEEE International Conference on Computer Communications*, San Francisco, CA, pp. 745–755, 2003.

37. M. J. Neely. Stochastic network optimization with application to communication and queueing systems. *Synthesis Lectures on Communication Networks*, 3(1):1–211, 2010.

38. R. Newton, S. Toledo, L. Girod, H. Balakrishnan, and S. Madden. Wishbone: Profile-based partitioning for sensornet applications. In *NSDI*, Boston, MA, Vol. 9, pp. 395–408, 2009.

39. A. Pathak, Y. C. Hu, M. Zhang, P. Bahl, and Y.M. Wang. Enabling automatic offloading of resource-intensive smartphone applications. Technical report, Purdue University, West Lafayette, IN, 2011.

40. J. M. Rabaey. *Digital Integrated Circuits*. Prentice Hall, Upper Saddle River, NJ, 1996.

41. R. Stanojevic and R. Shorten. Distributed dynamic speed scaling. In *Proceedings of IEEE International Conference on Computer Communications*, San Diego, CA, pp. 1–5, 2010.

42. E. Uysal-Biyikoglu, B. Prabhakar, and A. El Gamal. Energy efficient packet transmission over a wireless link. *IEEE/ACM Transactions on Networking*, 10:487–499, 2002.

43. A. Vetro and C. W. Chen. Rate-reduction transcoding design for wireless video streaming. In *2002 International Conference on Image Processing*, Rochester, New York, Vol. 1, pp. 29–32, 2002.

44. A. Wald. On cumulative sums of random variables. *The Annals of Mathematical Statistics*, 15(3):283–296, 1944.

45. S. Wang and S. Dey. Rendering adaptation to address communication and computation constraints in cloud mobile gaming.

In *Proceedings of Global Telecommunications Conference*, Miami, FL, pp. 1–6. IEEE, 2010.

46. Z. Wang and J. Crowcroft. Quality-of-service routing for supporting multimedia applications. *IEEE Journal on Selected Areas in Communications*, 14(7):1228–1234, 1996.

47. Y. G. Wen, X. Q. Zhu, J. P. C. Rodrigues, and C. W. Chen. Mobile cloud media: Reflections and outlook. *IEEE Transactions on Multimedia*, 16(4):885–902, 2014.

48. H. Wu, Q. Wang, and K. Wolter. Mobile healthcare systems with multi-cloud offloading. In *Proceedings of the 14th International Conference on Mobile Data Management*, Washington, DC, Vol. 2, pp. 188–193. IEEE, 2013.

49. M. Yang, Y. G. Wen, J. F. Cai, and F. C. Heng. Energy minimization via dynamic voltage scaling for real-time video encoding on mobile devices. In *Proceedings of IEEE International Conference on Communications*, Ottawa, Canada, pp. 2026–2031, 2012.

50. W. H. Yuan and K. Nahrstedt. Energy-efficient soft real-time CPU scheduling for mobile multimedia systems. In *Proceedings of ACM SOSP*, New York, pp. 149–163, 2003.

51. W. H. Yuan and K. Nahrstedt. Energy-efficient CPU scheduling for multimedia applications. *ACM Transactions on Computer Systems*, 24(3):292–331, 2006.

52. M. Zafer and E. Modiano. Minimum energy transmission over a wireless fading channel with packet deadlines. In *Proceedings of IEEE Conference on Decision and Control*, New York, pp. 1148–1155, 2007.

53. W. Zhang and Y. Wen. Cloud-assisted collaborative execution for mobile applications with general task topology. In *Proceedings of 2015 IEEE International Conference on Communication*, London, United Kingdom, 2015.

54. W. Zhang, Y. Wen, J. Cai, and D. O. Wu. Towards transcoding as a service in multimedia cloud: Energy-efficient job dispatching algorithm. *IEEE Transactions on Vehicular Technology*, 63(5):2002–2012, 2014.

55. W. Zhang, Y. Wen, and H.-H. Chen. Towards transcoding as a service: Energy-efficient offloading policy for green mobile cloud. *IEEE Network*, 28(6):67–73, 2014.

56. W. Zhang, Y. Wen, K. Guan, D. Kilper, H. Luo, and D. O. Wu. Energy-optimal mobile cloud computing under stochastic wireless channel. *IEEE Transactions on Wireless Communications*, 12(9):4569–4581, 2013.

57. W. Zhang, Y. Wen, and D. O. Wu. Collaborative task execution in mobile cloud computing under stochastic wireless channel. *IEEE Transactions on Wireless Communications*, 14(1):81–93, 2015.

58. W. W. Zhang, Y. G. Wen, Z. Z. Chen, and A. Khisti. QoE-driven cache management for http adaptive bit rate streaming over wireless networks. *IEEE Transactions on Multimedia*, 15(6):1431–1445, 2013.

59. X. W. Zhang, A. Kunjithapatham, S. Jeong, and S. Gibbs. Towards an elastic application model for augmenting the computing capabilities of mobile devices with cloud computing. *Mobile Networks and Applications*, 16(3):270–284, 2011.

CHAPTER 3

Design and Architecture of a Software Defined Proximity Cloud

Hyunseok Chang, Adiseshu Hari, Sarit Mukherjee, and T.V. Lakshman

Hyunseok Chang, Adiseshu Hari, Sarit Mukherjee, and T.V. Lakshman

CONTENTS

3.1   Introduction

Cloud computing has become the dominant mode of delivering Internet-based services to both fixed and mobile endpoints. Since 2008, most Internet traffic has originated or terminated in a data center [16]. Multiple factors have contributed to the rise of the cloud, including the widespread support for virtualization in the industry standard x86 processor architecture, a low cost VM rental model (Infrastructure as a Service or IaaS), the plunging cost of computer hardware together with the exponential increase in computer performance, high-bandwidth connectivity, and the rise of mobile devices and smartphones with the ability to stream various services from cloud-based back ends over the Internet.

As mobile devices improve in storage and processing power in accordance with Moore's law, services evolve in sophistication and complexity at the same rate, ensuring that the high-end mobile applications such as augmented reality or online gaming will continue to have a cloud-based back end. However, a cloud-based back end suffers from lag when the cloud is remote from end users and is not resilient to network and link failures. The inability of the OnLive [31]'s cloud-based gaming platform to dent the console gaming market shows that while the cloud can serve as a back end for general purpose apps, it is not suitable for high performance, low latency applications like interactive 2D or 3D gaming.

Beyond mobile applications, consider another class of applications that is becoming popular with the advent of various smart devices and the Internet of Things (IoT). We observe that a tremendous amount of data is generated at the edge of the network, for example, video captured by smartphones or tablets, machine-to-machine (M2M) sensor communications, surveillance and security video feeds from single-board computers (e.g., Raspberry Pi [32]), etc. Such edge-generated data is often transported back to the cloud for storage and processing, which requires a high-bandwidth connectivity into the central cloud. However, a majority of the data can be pre-processed and compacted. For example, user video can be compressed, M2M communication can be coded, and surveillance video can be filtered for abnormal incidents. With such pre-processing, the transport cost can be drastically reduced. To achieve this, however, large-scale computation is needed at the edge of the network, which is not supported by the centralized data center oriented cloud computing model.

What we observe from the cases earlier is that although a cloud back end is needed for these types of high performance applications, end users would benefit from lower latency and bandwidth consumption if the cloud back end were moved either wholly or partially as close to the users as possible. This requires a new model in which the centralized data center-based cloud is augmented with a presence at the network edge to support the emerging edge-based applications that are both computation and data intensive.

In this chapter, we present such a model of a hybrid cloud architecture we refer to as the *Proximity Cloud*, which is designed to deliver low-latency, bandwidth-efficient, and resilient end user services with a global footprint. Unlike a typical hybrid cloud that federate an enterprise cloud and a remote public cloud under separate management, the Proximity Cloud is an extended public cloud designed to bring the public cloud all the way to end users by leveraging compute nodes located at the enterprise or the home. This preserves and extends the standard IaaS-based virtual machine (VM) model of cloud computation by running VMs at the network edge in addition to the data center. The seamless interconnection of edge networks with data center networks enables latency-sensitive computation and user interaction components to run close to end users at the network edge, while additional heavy-duty processing and database components are hosted in the data center nodes.

In this architecture, end users benefit from the low latency between the users and the Proximity Cloud which is now part of their local networks. The latency benefit is particularly pronounced for users far from the data center. Another benefit is the bandwidth reduction provided by the pre-processing at the edge, which reduces the bandwidth requirements between the edge and the core.

The Proximity Cloud concept is highly synergistic with upcoming networking trends such as the emerging Network Functions Virtualization [28] architecture, the IoT, and the lightweight container-based virtualization. The Proximity Cloud allows typical middle box functionality like virus scanning, data compression, and application load balancing to be moved into standard compute nodes not only in the data center cloud but also at the edge, thereby optimizing bandwidth utilization on WAN links. The Proximity Cloud can function as an IoT gateway as it is able to access various edge resources such as sensors, laptops, and computers, which are not visible outside edge networks. New lightweight virtualization technologies like containers [23,27] allow us to create the Proximity Cloud for homes and small businesses using lightweight, low powered compute nodes like laptops and embedded devices.

The Proximity Cloud provides application resiliency in two ways. In case edge resources become temporarily unavailable, edge components of an application can fall back to the more reliable data center cloud. Conversely, if an edge component can provide an application's basic functionality by itself, the application can continue to function even without data center components (e.g., due to lost/flaky Internet connection or data center outage). In either case, the Proximity Cloud can provide graceful degradation of service—a particularly valuable feature for any cloud application.

Finally, the Proximity Cloud enables policy-based application and network design, in which enterprise policies can be used to dictate which data are allowed to move off premise into the public cloud, and which needs to be physically stored in the enterprise itself—a feature that is valuable for sectors like banking, which have critical data that needs to be retained in house, as well as for regulatory compliance with standards like HIPPAA and PCI DSS.

The Proximity Cloud model, while providing the many advantages listed earlier, raises new research challenges dealing with security, network address space translation, split cloud application models, and integration into existing cloud management frameworks. The focus of this chapter is the design and implementation of a prototype Proximity Cloud model that provides the listed features while solving the challenges listed earlier. This chapter describes the following aspects of the Proximity Cloud:

- A new paradigm for delivering services to the edge that combines the advantages of data center-based service delivery with local presence at the edge.

- An instantiation of this paradigm on the OpenStack cloud management platform with OpenStack extensions to support NAT-friendly and secure virtual tenant networks.

- Quantitative evaluation of the Proximity Cloud as well as two edge apps that we deploy in the Proximity Cloud: 3D indoor localization and video surveillance. We also describe a distributed Hadoop implementation in the Proximity Cloud.

3.1.1 Glossary

We use the following terminology to refer to the various components of the Proximity Cloud:

- *Edge network*: Edge networks are referred to as the part of the Internet to which end users connect directly, such as residential home networks, enterprise networks, or WiFi hotspots.

- *Edge zone*: An edge zone is defined as a collection of one or more dedicated compute nodes connected to a particular edge network.

- *Compute node provider (CNP)*: A CNP operates an edge zone, and offers one or more dedicated compute nodes in the zone for use/lease in the Proximity Cloud.

- *Proximity Cloud*: The Proximity Cloud is the federation of the data center compute nodes along with all the edge zones offered by different CNPs.

- *Proximity Cloud operator (PCO)*: A commercial entity that operates the Proximity Cloud out of data center resources as well as multiple edge zones.

- *Proximity Cloud central controller (PC3)*: A PC3 is a centralized management system for the PCO to manage the Proximity Cloud.

- *Edge apps*: Edge apps are end user services deployed in the Proximity Cloud by cloud tenants. An edge app is composed of a bundled set of IaaS images that are designed to launch and work together in unison in the data center and edge networks, to provide a service for mobile devices and smart IoT devices.

The rest of this chapter is organized as follows. We describe the Proximity Cloud architecture in Section 3.2, and its implementation on

OpenStack in Section 3.3. We present two functional edge apps as well as a Hadoop implementation in Section 3.4, and conclude in Section 3.5.

3.2   Proximity Cloud Architecture

The goal behind the Proximity Cloud architecture is to create a distributed hybrid cloud infrastructure, which utilizes data center resources as well as unused compute resources in edge networks in a seamless and secure manner. In the Proximity Cloud model, *edge networks* are part of the Internet where end users directly connect, such as home/enterprise networks or WiFi hotspots. An *edge zone* is defined as a collection of one or more dedicated compute nodes in a particular edge network, offered for use/lease by a single owner, referred to as a *CNP*. All compute nodes in the same edge zone are assumed to be part of the same, potentially NATed, network address space managed by its CNP. The *Proximity Cloud* is the federation of the data center nodes along with all the edge zones offered by different CNPs. The PCO is assumed to have a pre-existing, traditional IaaS data center cloud. By adding the Proximity Cloud functionality, the PCO can now extend the cloud's capabilities all the way to the edge networks close to end users. While edge zones may be offered by both individuals and enterprises, we believe that the likely usage scenarios involve zones offered by individuals being aggregated into the Proximity Cloud for public use, while the zones offered by enterprises will be reserved purely for the enterprise's own use, with the Proximity Cloud providing a managed cloud facility for the enterprise. Applications deployed in the Proximity Cloud, referred to as edge apps, are designed to launch and work together in unison in the data center and the edge nodes, to provide a service for mobile devices and smart IoT devices. Figure 3.1 shows the presence of the various edge zones in the Proximity Cloud.

3.2.1   Challenges in Building the Proximity Cloud

Before delving into the Proximity Cloud architecture, it is instructive to examine the differences between a standard cloud infrastructure and the Proximity Cloud to understand the challenges in building the Proximity Cloud.

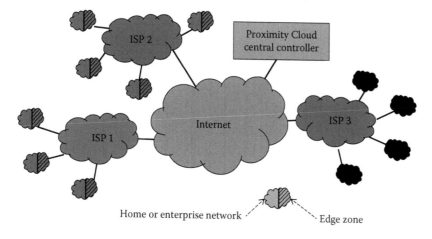

Figure 3.1   The distributed Proximity Cloud.

Heterogeneous compute and storage nodes: Unlike the fairly homogeneous nodes of a conventional data center, the edge compute nodes are extremely heterogeneous in their capabilities, and may lack many server class features. For example, they may not support 64-bit processing or hardware virtualization and are usually limited in their processor speed, memory, and disk storage capacity. Unlike conventional clouds where the smallest VM typically supports multi-GB RAM and disks of the order of hundreds of GBs (e.g., small instances in Amazon EC2), the limited capacity of the edge computing resources requires the ability to deploy *micro* VM images or lightweight containers, where required RAM is of the order of 128–512 MB, and disk storage is of the order of tens of GB.

Interconnect network: Unlike a conventional data center, there is no high speed, high bisection bandwidth interconnect between the various compute nodes in edge networks. While nodes within the same edge zones are likely to have a high bandwidth between themselves, the bandwidth across different zones is likely to be orders of magnitude less than in a conventional data center. Inter zone paths do not support native LAN or VLAN connectivity. Furthermore, the compute nodes in each zone are likely to be behind NAT gateways or restrictive zone-specific firewalls, complicating the task of interconnection.

Software resiliency and fault tolerance: Since the compute resources in the edge networks are likely to be older, they are also likely to exhibit failures at a much higher rate than conventional data center equipment. The Proximity Cloud therefore needs to provide resiliency by monitoring edge node failures and automatically restarting the services on failed nodes elsewhere, with minimal service disruption and network connectivity interruption. The Proximity Cloud can leverage its scale to its advantage to aggressively utilize the speculative execution features provided by cloud service and application frameworks such as MapReduce [22] or Hadoop [7].

Security: Unlike a conventional cloud where the compute nodes are under the physical control of the cloud operator, the Proximity Cloud relies on compute nodes provided by individuals and enterprises. This raises unique security challenges, such as the need to secure tenant VMs from malicious CNPs and vice versa.

Guided by the earlier points, our reference software driven Proximity Cloud management system, which we call the *PC3* is shown in Figure 3.2. The PC3 is designed to provide a complete networking and host virtualization management solution. One aspect of the PC3 design is that rather than providing all necessary functionality for the Proximity Cloud management from scratch, the PC3 relies on existing cloud management systems augmented with Proximity Cloud specific functionality. As shown in Figure 3.2, the PC3 is accessed via three different portals to reflect the three different players in the Proximity Cloud ecosystem. *Operator portal* is the administrative portal used by the PCO to manage the Proximity Cloud. *Tenant portal* is where cloud tenants log in to configure and manage their edge apps. A new portal, missing in traditional cloud management, is the *zone portal* which serves as a portal for CNPs to register their zones and associated compute nodes. The zone portal also provides CNPs with the initial software download necessary to integrate compute nodes into the Proximity Cloud.

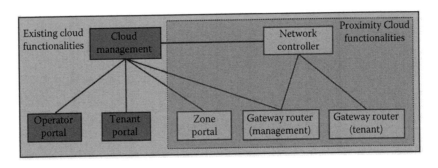

Figure 3.2   Proximity Cloud central controller (PC3).

Collocated with the PC3 are the gateway routers and the network controller, described in detail in Section 3.2.2. The management gateway router is used to bring the compute nodes located in various zones into virtual management networks accessible by the PC3. Similarly, the tenant gateway routers are used to bring tenant VMs provisioned in different zones into tenant-specific virtual networks. The network controller then provides an SDN-like controller to manage all these virtual networks in the Proximity Cloud. There is a management agent running on each compute node, which communicates with the PC3 to configure tenant networks, as well as the compute node's resources, such as starting and stopping edge app instances.

3.2.2 Networking in the Proximity Cloud

In a conventional cloud, the management traffic is carried over a secure connection between the cloud controller and each hypervisor. All VMs belonging to the same tenant are connected to a Data Link Layer (L2) bridge, which is either a software bridge in the hypervisor or a hardware switch accessed via a pass-through link connecting to the hypervisor. Tenant VM traffic that spans across compute nodes is relayed across bridges either using VLANs or via inter-bridge L3/L4 tunnels. The hardware and software switches providing the intra cloud network can be either traditional switches [2], or OpenFlow-based switches with a centralized SDN controller [13].

However, the traditional cloud networking model is not "edge friendly," in the sense that there is no framework for imposing per tenant intra-cloud bandwidth caps, and the tunneling mechanisms in use (e.g., VLAN-enabled switches or GRE-based tunnels) do not work across NATs. The key challenge for the network architecture of the Proximity Cloud is to overcome the following additional edge network-specific issues without any hardware support, while providing isolated per-tenant networks, which are fully transparent to tenant applications:

- Work across NATs.

- Fully isolate management and tenant traffic from each other and from any local traffic of the edge zones, while providing efficient, nontriangular routing among tenant application instances.

- Bandwidth management to ensure that the Proximity Cloud bandwidth usage does not eat into the edge zone's quota.

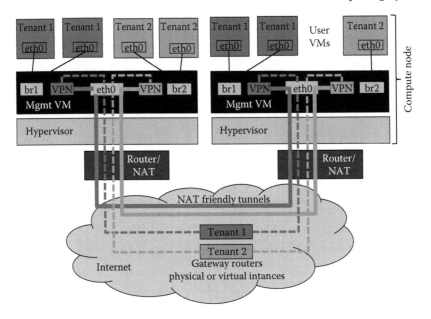

Figure 3.3   Networking in the Proximity Cloud.

- Work in an over the top mode without any managed networking components.

- Provide software resiliency by reassigning network configuration on demand from failed nodes to backup nodes.

We address the earlier issues using the model shown in Figure 3.3. We assume a model in which the compute nodes are dedicated to the Proximity Cloud and run Type 1 bare metal hypervisors. Each compute node runs a management agent on a standalone management VM (e.g., Xen Dom0). The principal networking components in the compute nodes are the centrally managed VPN and per tenant software bridges in the management VM, as shown in Figure 3.3.

Similar to the traditional approach, we rely on connecting the tenant VMs on a compute node to a bridge, which is a software bridge in the management VM since we do not rely on any hardware support. The key difference is that for the communication across tenant VMs, which reside on different compute nodes, instead of relying on the bridge's native packet relaying capability, we rely on off the shelf UDP/IP encapsulating, NAT traversal capable VPN software to connect the various bridges over VPN. We also use the same

VPN software to connect each compute node to the Proximity Cloud controller on the logical management plane network. Figure 3.3 illustrates these concepts. The figure shows two compute nodes in two different zones, each running VMs for two different tenants. All the VMs of one tenant are connected via one virtual network, while the VMs of the other tenant are connected to another virtual network. Not shown in Figure 3.3, but present in each compute node is another VPN connection to a management plane network, which connects all compute nodes to the Proximity Cloud controller via a management gateway router. Figure 3.3 also shows the two tenant gateway routers, which may be located either in the Proximity Cloud controller, or in the tenant premises, to carry all traffic entering and exiting the tenant network.

An important aspect of our solution is the configuration of the VPN software at each compute node. The VPN software needs to have enough information including necessary digital certificates to enable the node to route outgoing packets to a correct destination node, even if the destination node is sitting behind a NAT. Just like the way an SDN controller installs packet forwarding state in switches, the network controller installs the configuration information necessary for the proper functioning of the VPN software in each compute node and gateway router.

The overall effect of the SDN-like approach to tenant networking in the Proximity Cloud is to create separate, secure, failure resilient, logical L2 networks—one for management and one for each tenant. Note that tenant network address spaces are independent and, in case of private addresses, can be identical or overlap, and that the VPN-based tenant networking is fully transparent to the tenant VMs. While the management gateway router is collocated with the PC3, the tenant gateway routers can optionally be hosted by the tenant itself, either at the tenant's premise or at a separate data center. Off-loading the tenant gateway router from the Proximity Cloud cuts down on the bandwidth and processing capacity needed to set up and manage the PC3. The only requirement for the tenant gateway router is to have a public IP address and enough bandwidth to handle the network traffic to and from the gateway. In case the tenant gateway router is behind a NAT, it is necessary to enable port forwarding so that tenant VPN network traffic to a specified public IP address and port, control plane traffic from the network controller is directed toward the tenant gateway router.

Traditional VPNs are organized in a hub and spoke model, with all traffic between VPN clients flowing through a central VPN server. While this provides a working model, it is not optimal, particularly for the Proximity Cloud, since it is preferable for both latency and bandwidth reasons that traffic between two VMs flow directly, rather than via a central node. This type of functionality is provided by mesh networking VPN software. As a result, traffic between two tenant VMs on the same compute node, or two tenant VMs in the same zone will flow directly between them. In general, with mesh networking, traffic will flow directly between two nodes that are able to establish direct IP connectivity. In case the two nodes are behind separate NATs, then direct connectivity will depend on the nature of the NATs (e.g., cone vs. symmetric) and the effectiveness of the NAT traversal capability of the mesh VPN software. For those cases when NAT traversal fails, traffic between two nodes will be routed via the gateway router for that network. For example, in Figure 3.3, if it is possible to establish direct connectivity between the two compute nodes via mesh VPN software, then inter-node traffic will flow along the solid lines, while if direct connectivity is not possible, then traffic will flow along the dashed lines via the gateway router for that network. Bandwidth management can be provided by either static bandwidth allocation at each VPN endpoint, or via a dynamic bandwidth sharing scheme with distributed rate control as in [33].

The SDN-like control of standard VPN-based virtual networking in the Proximity Cloud provides a scalable solution to managing a large number of endpoints and networks while using off the shelf, well-tested software. An added advantage of using standard VPN software for the Proximity Cloud is that the tenant can also run the same VPN software at the tenant premises in order to connect to the tenant network, thereby allowing the tenant to directly connect to the network from a variety of fixed and mobile endpoints.

3.2.3   Zone Bandwidth Management

Edge networks have bandwidth caps imposed by the ISP they connect to. As a result, it is important to control the bandwidth usage by the compute nodes in each edge zone so that these bandwidth caps are not exceeded. Bandwidth management is complicated by the fact that there are possibly multiple nodes in a given zone, and that the traffic is generated by both management and tenant networks.

One option for bandwidth management is to route all traffic in a zone via a central choke point such as a designated compute node and implement bandwidth control there. However, this is inefficient unless the choke point is a gateway router. We assume a network design that is purely over the top without any managed network component such as home or enterprise edge routers.

It is possible to configure individual compute nodes with static quotas for each virtual network, but this is not efficient across all cases, especially when only some nodes are active. Our edge cloud architecture therefore relies on an approach with distributed rate control. Similar to the approach outlined in [33], this requires that each compute node within an edge network share its current bandwidth load with its peers so that each can work out their individual throttle factor for each virtual network, which is then applied to the VPN interface to each virtual network bridge. A central controller for the network helps manage this efficiently.

3.2.4   Edge Apps

As clarified earlier, we introduce the concept of edge apps that are designed to run in unison in the data center and the edge nodes, as end user services in the Proximity Cloud. An edge app is composed of two types of virtual networks and two sets of VM/container instances. One set of instances run in the data center, and the other set in the edge zone of the edge owner. In each set, there can be different types of IaaS images, and each type of image can be deployed on multiple instances (indicated by a replication factor). A virtual network called the *app-private network* is instantiated for each edge app on startup to interconnect all instances belonging to the app. The communication of the edge app with end users and resources in the edge network is provided by the *edge-local network*, which bridges edge app instances running in the edge nodes to the local edge network.

Figure 3.4 shows a sample edge app layout consisting of four IaaS images, labeled image1 through image4. image1 and image2 are used to realize the edge components with a replication factor of 1 and 2, respectively, while image3 and image4 realize the data center components with a replication factor of 3 and 4, respectively. image1 is connected to the edge-local network, while all components are connected to the app-private network.

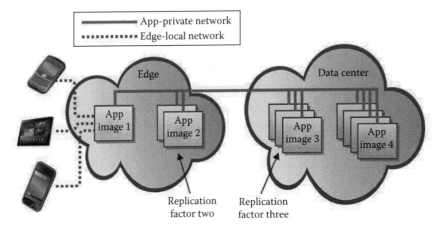

Figure 3.4   Anatomy of an edge app.

3.2.5   Security Aspects of the Proximity Cloud

In addition to security issues present in traditional data center-based clouds, the Proximity Cloud has unique, additional security issues since the compute nodes forming each edge zone are outside the physical control of the PCO, and share the same network as the other resources of the CNPs. We first note that the Proximity Cloud is *not* designed to run ultra secure applications, which need physical security, for example, payroll applications. Such applications may not be suitable even for a public cloud and are best done on the application user's own compute platform. However, even after discounting such applications, it is still necessary for the Proximity Cloud to provide at least a level of security suitable for commercialization of the Proximity Cloud concept. In this section, we discuss the various security issues raised by the Proximity Cloud, and how they can be addressed.

Protection from malicious cloud tenants: Can a malicious tenant damage the PCO, CNPs or other tenants? Note that protecting the PCO and CNPs from cloud tenants is a well-studied subject that also applies in the conventional cloud environment, and thus it is beyond the scope of this Proximity Cloud model described in this chapter (see, for example, [25] and references therein). In order to protect cloud tenants from one another, we provide network isolation via virtual networking as described earlier in Section 3.2.2.

Protection of CNPs from malicious CNPs: By participating in the Proximity Cloud, could CNPs actually jeopardize their own edge network and other connected devices, sharing the network? Such risk would be a huge disincentive for potential CNPs. To isolate individual CNPs from one another, we define zones on a per-CNP basis, and create a separate management network for each zone. Thus, a compute node owned by one CNP is prevented from directly accessing compute nodes owned by other CNPs.

Protection of PCO and cloud tenants from malicious CNPs: Unlike in the conventional data center clouds, the PCO and CNPs in the Proximity Cloud are *not* the same entity, and this raises more serious and unique security issues. Without adequate security, no tenant will run applications on the Proximity Cloud. To provide an adequate solution, we need to identify each vector through which a malicious CNP can attack the Proximity Cloud's management plane or tenant plane. These vectors are the following:

Live monitoring: A malicious CNP may offer virtualized instances of compute nodes to the PCO, and then monitor live compute nodes from the hypervisor underneath. In order to enforce only fully-dedicated physical machines as compute nodes, we prevent the installation of the Proximity Cloud software on VMs by detecting standard hypervisor signatures. Note that this cannot prevent cold boot attacks, in which a machine is powered off and the RAM is cooled in order to maintain its contents [1]. Again, we emphasize that applications that need physical level security should not run on the Proximity Cloud.

Console monitoring: The management console interface used by the hypervisor can potentially leak information that has possible security ramifications. While a typical management console interface requires secure login, it still needs hardening so that the CNP cannot make any changes to a running system or infer any information related to system activities. Standard practice for securing console access [37] or disabling any information-leaking console interface (e.g., XenServer's xsconsole) will go a long way in addressing this issue.

Network traffic sniffing: The management traffic is always encrypted. It is possible to encrypt the tenant traffic on a per-tenant basis, so that those tenants needing network security can choose to turn on encryption. Note that encryption comes with encapsulation/encryption overhead. This is studied in greater detail in Section 3.4.

Disk monitoring: It is possible for the CNP to examine or tamper with a disk after powering off a compute node. To prevent that, we add an encryption layer to compute node's disks, whose encryption keys are then remotely managed by the PCO. This setup protects cloud tenants from CNPs, giving the tenants an equivalent of data center-based security. Cloud tenants can choose to add another layer of encryption from within their VMs (using their own encryption keys) if they want to protect themselves from the PCO, much like Amazon EC2 customers encrypt EBS volumes.

Denial of Service: A malicious CNP can prevent a cloud tenant from completing his task by frequently powering off the provider machines, or disconnecting them from the network. An eBay-like rating system can help identify good and bad CNPs. Like eBay, the PCO works purely in a matchmaking capacity by bringing together CNPs and tenants. Like eBay, one important goal is to identify fraudulent sellers (providers). Outside the rating system, we can also have requirements for reliability and uptime as measured by periodic pings from the PC3. Those CNPs who do not meet the operator SLA can be disadvantaged or barred.

3.3   Proximity Cloud Implementation

The goal behind our Proximity Cloud implementation is to realize the Proximity Cloud by using as many off the shelf software components as possible, while providing customization for building differentiating features of the Proximity Cloud. While there can be more than one approach to realizing the Proximity Cloud architecture elaborated in Section 3.2, our implementation of the Proximity Cloud is mostly driven by the underlying Proximity Cloud environment.

We implement the Proximity Cloud architecture by using OpenStack, an industry-standard cloud management platform. Since OpenStack is originally designed as an IaaS platform for data centers, there are several challenges in building the Proximity Cloud on top of OpenStack. For one, the Proximity Cloud needs to support low cost, low power hardware platforms which are not ideal to run a full-blown hypervisor due to their limited CPU and memory resources, or lack of hardware virtualization support. Besides, unlike in the well-guarded data center environment, edge nodes may be exposed to potentially malicious CNPs, which necessitates an additional layer of protection for any edge apps running on the edge nodes. Finally, there is no mechanism in OpenStack for managing or deploying edge apps on

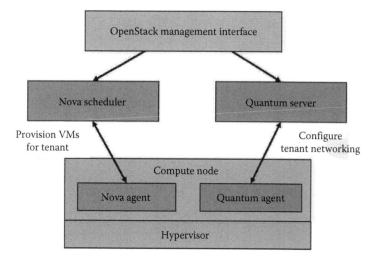

Figure 3.5   OpenStack architecture.

compute nodes which are spread across heterogeneous, potentially NATed, L3 networks. In the following, we describe how we address these challenges in the prototype implementation (Figure 3.5).

3.3.1   Compute Node Management in the Proximity Cloud

OpenStack comes with support for multiple hypervisors such as Xen, KVM, vSphere, as well as lightweight Linux containers such as LXC or Docker. Despite its multi-hypervisor/container support, OpenStack does not allow one to use more than one type of hypervisors or containers technologies on each compute node. We extended OpenStack Nova [29] so that it can deploy a mix of hypervisor VMs and LXC containers on the same compute node. That way, we can utilize resource-limited edge nodes more effectively to host various edge apps with different requirements. For example, edge apps which have incompatibility with the underlying host operating system can be deployed in full hardware-based virtualization environment, while most other Linux-native apps can be deployed in lightweight containers with minimal overhead.

As pointed out earlier, we enable encryption of disk storage of compute nodes as a protection mechanism against potentially malicious CNPs in the Proximity Cloud. We encrypt compute node's raw disk drives by using dm-crypt, which provides transparent encryption of block devices using the device-mapper interface of the Linux kernel. The disc encryption keys used by Linux Unified Key Setup (LUKS) are created and managed by the PCO. Using dm-crypt and LUKS, the PCO can remotely manage encryption keys for compute node storage, and securely unlock encrypted root partitions over SSH during compute node reboots [15].

3.3.2   L2 Tenant Networking in the Proximity Cloud

In OpenStack Folsom [24], Quantum component is dedicated to managing virtual tenant networking. We extend OpenStack Quantum to support the virtual networking in the Proximity Cloud. In our extended Quantum implementation, Quantum server and agent communicate via a custom Quantum plugin to manage virtual networks for different edge apps on demand. In the rest of the implementation section, we describe this extension in more detail.

3.3.2.1   Proximity Cloud Quantum Plugin

Quantum can support various network virtualization technologies by leveraging the notion of "plugins." We design and implement a custom Quantum plugin that allows OpenStack controller to manage edge nodes located behind NATs across different L3 networks, and to instantiate virtual app-private networks spanning these nodes. First, we assume that there is a dedicated management network over VPN between Quantum server (i.e., server component of Quantum) and Quantum agent running on each edge node. Such VPN can easily be created by Quantum agent even behind NAT. Quantum agent is then responsible for configuring app-private networks across edge nodes. Since an app-private network may span across multiple NATed edge nodes over different L3 networks, Quantum agent creates per-app virtual L2 underlay that interconnects edge nodes over physical L3 networks, over which actual app-private networks are maintained. Per-app underlay configuration information is centrally managed by Quantum server, and shared among compute nodes involved to allow connectivity among different edge app instances over an underlay.

3.3.2.2   VPN Underlay for L2 Tenant Networking

To create NAT-friendly L2 underlay for app-private networks, we use the open-source tinc VPN software [14]. Using VPNs for L2 underlays has several advantages in the Proximity Cloud. First, it allows edge nodes residing in different, potentially NATed, L3 networks to be integrated into the address space of the data center. Second, it provides security by encrypting the communications to and from the edge nodes, which is important because the edge nodes are no longer in a trusted zone such as the data center. Third, it provides a single point of control for all the Proximity Cloud traffic to and from each edge node. This point of control can be used to monitor and optionally throttle the Proximity Cloud traffic.

Our choice of tinc VPN software is driven by the fact that it supports mesh-based VPN topologies. In a mesh-based VPN, any node can initiate/serve a connection to/from any other node, and form a direct connection in between. Furthermore, tinc has STUN [34] functionality built in, and any VPN node with publicly reachable IP address can play the role of a STUN server, bootstrapping nodes behind cone-type NAT gateways to connect with each other directly. This means that traffic from edge app instances running on different NATed nodes will flow directly between the two nodes without any intermediary node, which greatly improves inter-NAT delay performance and removes any scaling bottleneck in L2 underlays.

3.3.2.3   Extended Quantum Server and Database

As described before, Quantum server stores and manages per-app VPN configuration information including public keys for authentication. To support this, we extend Quantum database schema by adding a new table called "app public key." Each entry in the table stores a public key associated with a specific compute node, generated for an edge app which has any instance on the node. We then extend Quantum RESTful APIs and RPC communication supported by Quantum server, so that a given app's public key can be added or removed in Quantum database, and the public key table can be looked up using several search criteria (e.g., App ID, network ID).

3.3.2.4   Public Key Server

To set up a mesh-like per-app L2 underlay among a set of compute nodes over VPN, each participating node must know the public keys of all the other nodes in the underlay. To ensure that app-specific public keys are properly deployed on all participating nodes, we rely on a separate "public key server." The role of the public key server is to distribute or reclaim public keys to and from the compute nodes involved in the app-specific underlay, upon any change in node membership of the underlay. We develop a simple public key server which deploys or removes public keys on compute nodes over SSH, upon requests from Quantum agent. By design, our public key server is purely stateless; public keys are not stored inside the key server, but rather retrieved from Quantum server via Quantum APIs. That way, there is less administrative overhead in operating the key server, and no inconsistency between the Quantum server and the public key server.

3.3.2.5 Quantum Agent

Quantum agent is a daemon process running on each compute node. The agent communicates with Quantum server via Quantum APIs over a RPC connection, and uploads/downloads app-specific public keys to/from Quantum database. Upon request from Quantum server to create a new app-private network and to add a particular edge app instance to the network, Quantum agent proceeds as follows:

- If there is no L2 underlay provisioned for the app-private network on this compute node, generate a private/public key pair for the app-private network on the node.

- Upload the created public key to Quantum server via RPC, which is stored in Quantum database.

- Request a public key server to distribute the public key to other compute nodes connected to the app-private network.

- Retrieve from Quantum database all available public keys that have been deployed on other compute nodes for the same app-private network, and VPN port number to use for the app-private network.

- Using collected public keys and port number, generate necessary underlay configuration, and launch an L2 underlay associated with the app-private network.

- Finally, attach the underlay to the app-private network (i.e., add the underlay's VPN interface to app's Linux bridge).

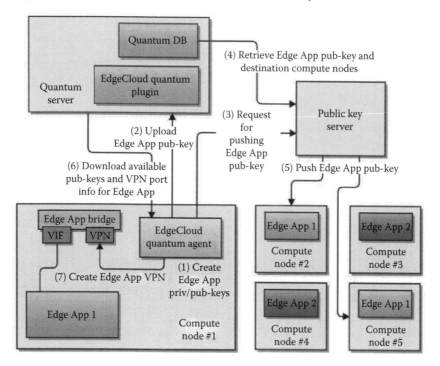

Figure 3.6   Quantum extension in the Proximity Cloud.

The step-by-step process of provisioning an app-private network via the Quantum plugin/agent is illustrated in Figure 3.6.

3.3.3   Zone Bandwidth Management

For the proof-of-concept prototype implementation, we rely on a very simple static bandwidth allocation for edge zones, where inbound and outbound bandwidth for each zone is statically defined. While standard hypervisors provide primitives for statically configuring bandwidth caps of individual virtual interfaces, enforcing bandwidth caps on a virtual interface level is not desirable since that will unduly rate limit traffic between VMs hosted within the same zone. On the other hand, defining static bandwidth caps on a compute node basis is not efficient either, especially in a multi-node zone, since idle bandwidth of nonactive compute nodes will not be properly utilized by active compute nodes. As a compromise, we rely on the NAT gateway's rate limit functionality, thereby enforcing per-zone bandwidth allocation.

In a sense, we mimic the net effect of dynamic bandwidth allocation as sketched in Section 3.2.3, in a rudimentary fashion.

3.4 Evaluation

In this section, we validate the feasibility of the infrastructure cloud services enabled by the Proximity Cloud. We also show the implementation of several edge apps, and present experimental results. In the experiments, we focus on illustrating the benefits of the Proximity Cloud as an end user services delivery platform, for example, reducing latency and bandwidth of application traffic and providing graceful degradation of service.

3.4.1 Infrastructural Overhead of Security

Compared to the conventional data centers, the Proximity Cloud incurs additional overhead for managing compute resources, mainly for security reasons. In the Proximity Cloud, all data communications between different edge zones are encapsulated and secured via VPN underlays. Every VPN packet incurs overhead from the encapsulation and encryption/decryption process, and this overhead takes a toll on network throughput and CPU resources of compute nodes. Furthermore, if we enable disk encryption on compute nodes in order to protect against malicious CNPs tempering with compute node's physical disk drives, this will be another limiting factor for CPU resources left available to tenant VMs. We evaluate the implication of these protection mechanisms in terms of network throughput and CPU resources.

First, we evaluate the overhead of VPN by conducting the following experiment. We interconnect two compute nodes using an Ethernet cable over a rate-limiting dummynet [5] based network emulator. We then set up a tinc VPN session between the two, and run iperf to exchange traffic between them over VPN. The rate-limiting dummynet simulates the link bandwidth between the two nodes. In this setting, as we increase link bandwidth between the two compute nodes, we check the resulting VPN throughput and tinc's CPU utilization. The compute nodes used in the experiment are commodity laptops with dual CPUs and 3GB memory, which will be typical hardware resources available in edge zones.

In Figure 3.7a, we plot two quantities as a function of link bandwidth: (1) ratio of VPN throughput and link bandwidth (denoted as

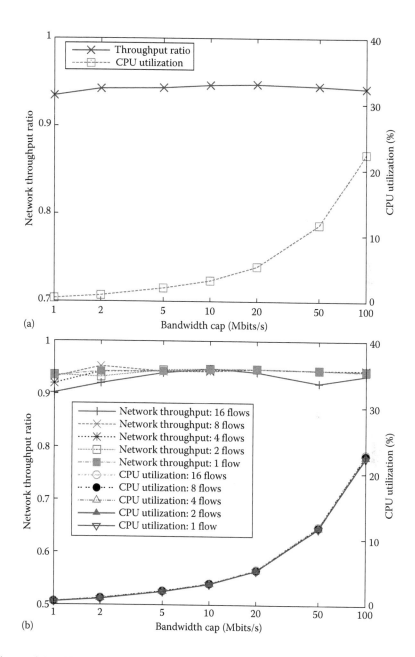

Figure 3.7 VPN overhead. (a) CPU and network throughput and (b) effect of multiple flows.

network throughput ratio) and (2) tinc's CPU utilization. The network throughput ratio remains around at 0.93 over a wide range of link bandwidth. About 7% of encapsulation overhead is to be expected when VPN header size and MTU are considered. On the other hand, tinc's CPU utilization increases linearly with increasing link speed, which is no surprise due to the fixed per-byte VPN processing CPU overhead. It is worth noting that when the link speed is in the range of typical residential broadband connections (e.g., less than 20 Mbits/s [12]), the VPN's CPU overhead on this particular compute node remains less than 5%. The effect of the number of traffic flows on VPN overhead is negligible, as shown in Figure 3.7b.

Next, we perform another experiment to examine the overhead of disk encryption. When WRITE/READ access is performed on a disk encrypted by dm-crypt, a kernel process called kcryptd performs the cryptographic routine of dm-crypt. Thus, the CPU overhead of disk encryption can be measured by the CPU utilization of kcryptd process. In this experiment, we prepare disk partitions encrypted by dm-crypt on both internal hard drive and USB hard drive, and perform disk WRITE and READ operations with them, using dd command with direct I/O option. We adjust the CPU allocation of kcryptd process during WRITE/READ access, and check the resulting I/O throughput. We control the CPU usage of kcryptd process by running other computationally intensive jobs (e.g., infinite while loop) concurrently, and running the kcryptd process with different priorities.

Figure 3.8 shows the relationship between I/O throughput and cryptographic CPU overhead for two different storage media. Marks are included in the figure to show the maximum achievable I/O throughput of encrypted storage media. For example, with an encrypted USB hard drive, maximum achievable throughput of WRITE operations is 28 MBytes/s.

Given the same CPU utilization, there is little difference in I/O throughput between internal drives and USB drives, except that the maximum I/O throughput achieved in internal hard drive is marginally higher than in a USB hard drive. While accessing an encrypted drive at full speed will result in 20%–25% CPU overhead on this particular laptop, we would like to point out that the implication of this result varies depending on which application is running on the Proximity Cloud. For cloud backup type applications, the bottleneck of backup operations will be at the network, where network throughput will typically be an order of magnitude lower

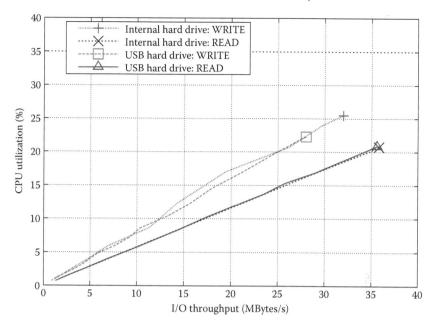

Figure 3.8   Disk encryption overhead.

than maximum achievable I/O throughput. In such an environment, the CPU overhead of I/O access will not be significant. On the other hand, for MapReduce type applications where internal disk access will prevail, we will need to carefully rate control disk I/O of tenant VMs, in order to supply adequate CPU resources to tenant VMs.* A point worth noting is that due to the rising importance of security, recent generations of Intel processors (e.g., Core i5 and i7) come with embedded hardware support for encryption and decryption, such as AES-NI [19]. In addition, Intel publishes a suite of software libraries (e.g., Intel IPP [8]) that are highly optimized to take advantage of AES-NI. OpenSSH used by t inc VPN for packet encryption, as well as disk encryption reportedly benefits greatly from such hardware and software supports offered by processor makers. For example, Intel's performance testing reports over 70% acceleration of disc encryption with AES-NI [19], and 3–10-fold speed boost of OpenSSL with AES-NI and IPP [38], compared to non-AES-NI. Once processors with

---

* One approach to control I/O bandwidth on per-VM basis is to use dm-ioband device-mapper driver [36] which can be installed in a compute node's management VM to enforce per-VM I/O bandwidth management policies.

AES-NI support become more popular in mainstream desktops and laptops, that will greatly improve the performance of the Proximity Cloud.

3.4.2   Edge App Evaluation

To demonstrate the benefit of the Proximity Cloud, we present the implementation of three edge apps, and evaluate their performance when deployed in the Proximity Cloud.

3.4.2.1   3D Indoor Localization Application

The first edge app we present is 3D indoor localization, where the goal is to find the 3D position and orientation of a user's mobile device in a particular indoor environment. The high-fidelity 3D localization increasingly finds practical usage in augmented reality and interactive guidance applications.

In [39], we present a vision-based 3D indoor localization system for smart devices such as smartphones or tablets, which can achieve sub-meter level 3D localization accuracy. In this approach, an indoor map that captures the 3D geometry of a given indoor environment is pre-built by using depth-sensing camera [10]. In order to find the 3D position and orientation of a smart device, we first capture a 2D photo image and accelerometer measurement from the device. Based on this data, we then compute the translation vector and rotation matrix between 3D indoor map and the viewpoint captured in 2D photo, by solving a nonlinear optimization problem with six variables (i.e., three-axis position/orientation of a rigid body) iteratively. It turns out that this optimization problem is too computation-intensive to run natively on the current generation of smart devices, not to mention the storage and bandwidth needed to download and store 3D indoor maps. This makes 3D indoor localization an ideal edge app that can leverage low-latency compute resources at the edge.

In the following, we describe how we deploy the 3D indoor localization app in the Proximity Cloud, and then present experimental results relevant to the Proximity Cloud environment.

3D Indoor Localization in the Proximity Cloud

The cloud portion of the 3D indoor localization is powered by JPPF parallel computing framework [9] and is spread across the Proximity Cloud and an external data center. When the connectivity to

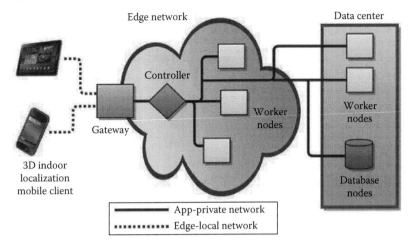

Figure 3.9   3D indoor localization app.

the data center goes down, the indoor localization app can continue to function on the Proximity Cloud, possibly with degraded performance—the type of application resiliency mentioned before. The cloud infrastructure needed for the app consists of four components as shown in Figure 3.9.

Gateway: It is the front end of the 3D indoor localization app which exposes a web service interface to end users. The gateway and end users communicate with each other via an edge-local network. The localization client software running on user's device captures a 2D photo and accelerometer measurement, and sends a localization request to the gateway. When the gateway receives the request, it delegates localization computation to a compute cluster via an app-private network, and sends the result back to end user after computation is done.

Controller: The controller is a job scheduler component of the compute cluster, responsible for coordinating the overall execution of localization computation in the compute cluster. Controller receives a job from the gateway, splits the job into multiple tasks, and assigns the tasks to individual compute nodes in the cluster for parallel execution. Each task is then run independently, and its output is sent back to the controller. Controller aggregates the output of individual tasks, and sends a final result to the gateway.

Worker node: A worker node executes tasks assigned by the controller. It can be deployed at the edge (for latency) or in the data center (for reliability). Different worker nodes and the controller are inter-connected with an app-private network.

Database node: A database node is an optional node in the data center that physically hosts indoor map database files for worker nodes to access during localization computation. Alternatively, map database files can be co-located on each worker node, in which case the database node is not necessary.

The deployment of 3D indoor localization app in the Proximity Cloud involves creating an app-private network dedicated to the app, and installing three app images and one database on the afore-mentioned four components. The app image for the gateway contains a Java Servlet for indoor localization, while the app images for controller and worker nodes contain the JPPF framework components. A database node houses and exports (e.g., via NFS) indoor database files.

Latency Experiments

The full-functional 3D indoor localization app is deployed on the Proximity Cloud, which consists of 12 node instances (1 gateway, 1 controller, and 10 worker nodes). The 45 MB indoor map, which captures the 3D geometry of an office room of 3 m × 3 m × 3 m dimension, is deployed on each worker node. We use pre-recorded 35 camera shots of the office room and accelerometer measurements, both taken from Galaxy tablet PC, to conduct 3D localization. We use netem, kernel support for network emulation in Linux, to emulate different network latencies between nodes. The metric used in our evaluation is the user-perceived end-to-end delay of 3D localization (i.e., time taken by client to receive a result after requesting for localization). In our experiments, we consider several deployment scenarios for 3D indoor localization app.

Deployment in the data center. We first examine the deployment scenario where the app runs purely at the edge or in the data center. To emulate the data center deployment, we vary the network latency between client and the gateway, and examine what impact it has on the end-to-end 3D localization delay. The result is shown in Figure 3.10a. The x-axis represents the average network latency to the data center. The special case of zero RTT delay corresponds to the deployment purely at the edge. The figure shows that the 3D localization delay increases linearly with network latency to the data center.

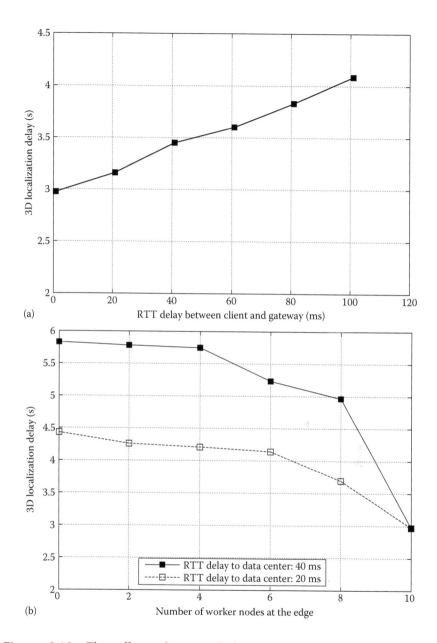

(a)

(b)

Figure 3.10 The effect of network latencies on 3D indoor localization delay. Latencies between (a) client and gateway, (b) controller and workers. (*Continued*)

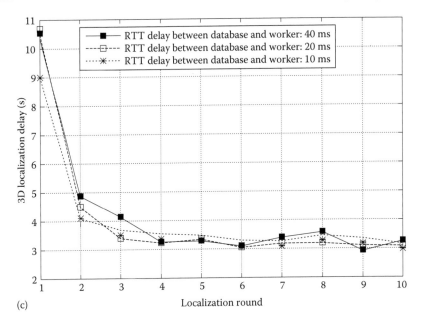

(c)

Figure 3.10 (*Continued*)   The effect of network latencies on 3D indoor localization delay. Latencies between (c) workers and storage.

An increase of 100 ms latency leads to 25% higher 3D localization delay. This illustrates the importance of cloud proximity in latency-sensitive, computation-intensive cloud apps such as 3D localization. More generally, the curves in Figure 3.10a could be pushed up or down depending on the complexity of computations needed for a given app, but the slope of the curve would remain the same due to network latency constraints.

Hybrid deployment. Next we consider a hybrid deployment where 3D indoor localization is deployed partially at the edge as well as in the data center. For the experiment, we first configure the network latency between the controller and 10 worker nodes to a small value (e.g., 0.5 ms), as if the worker nodes were located in the local edge network. Then we choose some worker nodes out of 10, and artificially increase their network latency from the controller, to a set value (e.g., 20 or 40 ms), as if the close-by edge nodes were replaced by distant data center nodes. As we move more worker nodes from the edge to an external data center like this, we check how that affects user-perceived 3D indoor localization delay.

In Figure 3.10b, we plot the 3D localization delay as a function of the number of worker nodes at the edge. The left extreme on x-axis corresponds to a case where all 10 worker nodes are in the data center, while the right extreme means all 10 worker nodes at the edge. The figure demonstrates the degraded performance of indoor localization computation (in terms of delay) as edge nodes start to disappear. In this particular app, the increase of 3D localization delay with fewer edge nodes is substantial. The reason for such sensitivity is that the 3D localization computation involves *multiple* rounds of interactions between the controller and worker nodes to solve smaller subproblems (e.g., feature point matching, iterative optimization for localization). Thus, even a small increase in network latency between the controller and worker nodes leads to a much longer end-to-end localization delay. While this observation only applies to a particular implementation and setup of our 3D indoor localization app, the benefit of edge processing and graceful degradation is still valid.

Deployment at the edge with storage in the data center. Next, we examine the effect of the storage location in the deployment. For 3D indoor localization, storage is needed to store indoor database map files. While indoor map could be deployed on individual worker nodes along with application binaries, multiple compute nodes could share storage for space efficiency or low app provisioning time. In this experiment, we provision a separate database node which houses and exports locally stored indoor map files to other worker nodes via NFS. To emulate a database node in the data center, we vary network latencies between the database node and worker nodes. On each worker node, we enable file caching for remote NFS share, so that it can locally cache previously accessed indoor map files. In this setup, client continues to issue localization requests, a batch of 10 requests in each round. We examine average 3D localization delay in each round as the number of rounds increases under varying network latencies, and plot the result in Figure 3.10c. The figure shows the dramatic performance difference between "cold" cache (e.g., round 1–2) and "hot" cache (e.g., round 4–), across all network latencies. The dramatic increase in localization delay with cold cache demonstrates the importance of storage proximity for compute nodes. For the 3D localization application which typically runs over a long term with the same data sets, the storage latency issue can be mitigated by file caching as shown, when the database node is provisioned in a distant data center.

3.4.2.2   Video Surveillance Application

The second edge app we present is Scene Activity Vector (SAV) based video monitoring and surveillance [30], which is based on a highly-compressible video feature and motion detection and filtering algorithm. Due to the popularity of single-board computers (e.g., Raspberry Pi) nowadays, IP camera-based surveillance becomes increasingly popular. Video surveillance applications are usually based on monitoring and archiving video feeds collected from IP cameras at a central data center. In comparison, the SAV-based video surveillance application can optimize the bandwidth consumption of video surveillance feeds by highly compressing the feeds between the edge and the central site. The input to SAV is a sequence of uncompressed motion video images obtained from a video stream, and its output is a series of SAVs. Compared to the original video stream from the IP camera, SAV is highly compressed, with compression ratios exceeding 100 or more.

Our edge app implementation of the SAV-based video surveillance App is as shown in Figure 3.11. There are two instances—one in the edge and the other in the data center. The IP camera is mounted on Raspberry Pi. The SAV module is implemented as a gstreamer [6] module. The gstreamer framework is used to create a three-way

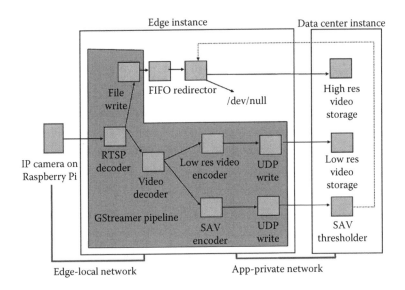

Figure 3.11   SAV-based video surveillance app.

pipeline as shown in the figure. The first pipeline takes the original high-resolution video stream and sends it to a FIFO pipe. The subsequent Redirector pipe is configured to discard it by default, but can also send it to a data center instance on demand. The second pipeline decompresses the video, and sends it to SAV processing before sending it upstream. The third pipeline decompresses the video before re-compressing it to a lower resolution and framerate, and then sends it upstream. The data center instance consists of an archiving module which takes the low-resolution video for archiving, and a SAV thresholding module which generates a trigger when the SAV activity exceeds a threshold. When a trigger is generated, the Redirector pipe of the edge node is re-configured to send the data upstream for additional processing. After a specified time period, the Redirector pipe reverts to its original setting.

Figure 3.12 shows the bandwidth consumption of the incoming RTSP video stream from IP camera, the output SAV stream and the output low-resolution video stream over a representative interval. The IP camera is configured to generate a H.264 stream at 15 fps

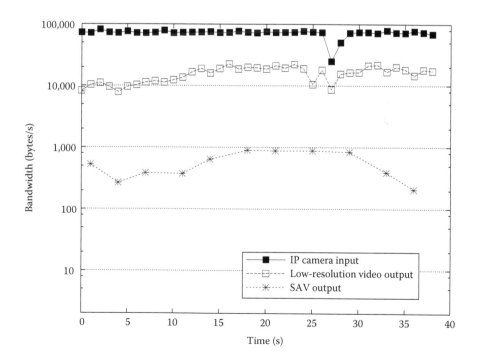

Figure 3.12   Bandwidth usage: SAV vs. IP camera.

with 640×480 resolution. The low-resolution video is a 340×240 FLV stream. Over the time interval shown in the graph, the bandwidth of SAV was merely 0.2% of the IP camera, and the low-resolution video consumed 21% bandwidth needed by the IP camera, which illustrates the fact that edge apps can be used to provide bandwidth optimized services at the edge by leveraging edge resources.

Another benefit to edge owners is graceful degradation or service resiliency. It is possible to run the SAV thresholding program in the edge instance itself, rather than in the data center, at the cost of reduced accuracy, as the data center instance can run on faster hardware, and access historical databases for more intelligent surveillance. In case of network outage, video surveillance can still continue to run by switching to the thresholding module running in the edge instance.

3.4.2.3   HDFS/Hadoop Application

Finally, we deploy a representative data center application on the Proximity Cloud, and evaluate its performance. A popular class of applications run in today's data centers is MapReduce-driven Hadoop applications [7], and so it is natural to ask how such computation and I/O intensive applications would perform in this new type of cloud architecture. As clarified already, one key aspect of the Proximity Cloud is the distributed nature of compute resources in heterogeneous edge networks with limited bandwidth. The implication is that the performance of deployed applications would be affected to a great extent by where and how application instances are deployed and scheduled within the Proximity Cloud. Placement and scheduling will be a less of a concern in the traditional data center-based cloud, where plentiful server horsepower and high bandwidth interconnects are available. We investigate the effect of the Proximity Cloud specific factors on Hadoop performance.

To start with, we set up a small Proximity Cloud testbed in a virtualized environment, consisting of VM instances of compute nodes and NAT gateways, created on multiple servers. We construct each edge zone consisting of one NAT gateway and multiple compute nodes connected behind it. In order to conduct experiments in a bandwidth-controlled environment, we specify a QoS traffic-shaper policy on all NAT gateways, so that both inbound and outbound traffic on each zone is rate-limited. We assign to each zone uplink/downlink bandwidth proportional to the number of compute nodes in the zone.

Table 3.1 Hadoop Testbed Configuration

Zone	Zone 1	Zone 2	Zone 3
Uplink bandwidth (Mbits/s)	3	2	3
Downlink bandwidth (Mbits/s)	6	4	6
Number of compute nodes	3	2	3
Number of cores per node	2	2	2
CPU type (bit)	64	64	64
CPU speed per core (GHz)	2.40	2.40	2.93
Memory per node (GB)	4	4	4

The testbed consists of eight compute nodes distributed across three different zones. The detailed zone configuration including hardware specification of compute nodes is described in Table 3.1.

In the testbed so constructed, we deploy eight VMs belonging to a hypothetical tenant, each of which has 1 GHz single-core CPU and 1 GB memory allocated. We choose the same number of VMs as there are compute nodes in our prototype, so that there is no more than one VM for the tenant on each compute node. Deploying each VM on a distinct compute node is for reliability at the cost of increased inter-host traffic. These VMs are on the same L2 tenant network dedicated to this particular tenant. We prepare a separate tenant gateway router attached to the tenant network, serving as the default gateway for the tenant network. By using Cloudera CDH [3], we then set up a Hadoop cluster consisting of eight tenant VMs and the tenant gateway router. In Hadoop, there are four principle entities: NameNode (HDFS metadata server), DataNode (HDFS data storage), JobTracker (job scheduler), and TaskTracker (map/reduce worker node). We designate the tenant gateway router as NameNode and JobTracker, each of all eight VMs as TaskTracker, and one VM in each zone as DataNode, serving as HDFS storage for the zone. So there are three VMs, one in each zone, assigned dual role of TaskTracker and DataNode. The rest of VMs run as TaskTracker only. We run sample Hadoop applications on this Hadoop cluster, and report the result in the following. The HDFS/Hadoop configuration used is shown in Table 3.2.

Table 3.2 HDFS/Hadoop Configuration

dfs.replication	dfs.block.size	mapred.max.split.size
3	128 MB	16 MB

TeraSort application sorts a large number of 100-byte randomly generated records, and as such it consumes a considerable amount of Hadoop resources in the form of computation, networking, and storage I/O access. We first generate one million records, and store them in the HDFS-backed storage with default replication factor three. Since we have three DataNodes in the HDFS storage (one DataNode in each zone), it is ensured that each zone has a copy of one million records in its designated DataNode. In this setting, we run TeraSort application on the stored records.

In Figure 3.13a, we visualize the progress of TeraSort job execution with various map and reduce task assignment. The y-axis of the figure plots the progress level of TeraSort job in percentage, and the elapsed time to reach a particular progress level is indicated on the x-axis. Like any Hadoop application, TeraSort goes through a sequence of map and reduce steps. In the figure, the range of 0%–50% on the y-axis captures the initial map stage, and the range of 50%–100% corresponds to the subsequent reduce stage. As we increase the number of reduce tasks while using a fixed number of map tasks (e.g., 8), there is a sweet spot in the middle (at three reduce tasks) where job completion time is minimized. This behavior can be explained as following. When JobTracker assigns reduce tasks to available TaskTracker nodes, it takes data locality into consideration. In case of three reduce tasks, JobTracker will assign them to three DataNodes available in the Hadoop cluster, so that the reduce tasks can store final results on to locally available HDFS storage. On the one hand, if we create more reduce tasks than there are DataNodes, there is chance that reduce tasks are assigned to TaskTracker nodes with no local HDFS storage attached, and thus they need to transfer reduce output onto a separate DataNode, slowing down the reduce stage. On the other hand, creating only one or two reduce tasks consumes even more expensive inter-zone bandwidth resource because reduce tasks now need to fetch the intermediate map output from most of other TrackerNodes' local storage, creating inter-zone traffic. As for the effect of map task assignment, increasing the number of map tasks from three to eight helps with the overall job completion time by close to half. While this result seems obvious, we would like to point out that increasing map task count helps with job completion only when there are enough DataNodes. Otherwise, more map tasks may actually slow down job performance by creating additional inter-VM and inter-zone traffic.

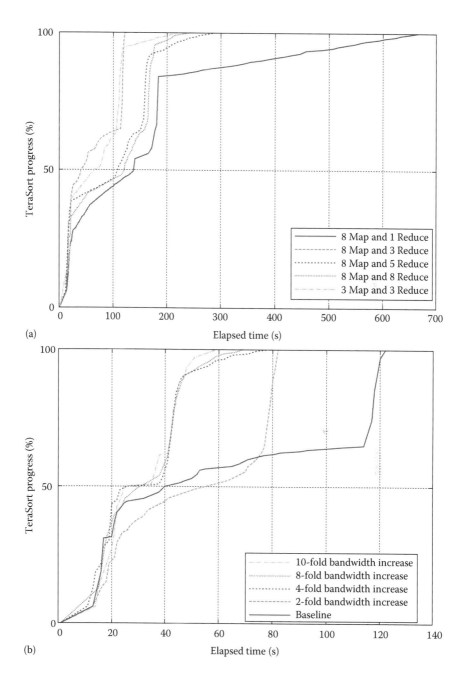

Figure 3.13   TeraSort performance comparison. (a) Effect of map/reduce task allocation and (b) Effect of network bandwidth.

Figure 3.13a suggests that network bandwidth plays a particularly important role in MapReduce performance in the Proximity Cloud environment. In Figure 3.13b, we verify this conjecture by actually controlling bandwidth rate limit in our testbed. We scale up uplink/downlink bandwidth limitations of our testbed gradually, and repeat our TeraSort experiment. The figure shows that with the increase of network bandwidth, the performance boost of TeraSort tends to be marginalized. This finding suggests that while network bandwidth is a dominant factor for MapReduce performance in a low bandwidth environment, it becomes less of a bottleneck for MapReduce jobs in higher bandwidth settings.

3.5   Conclusions and Related Work

In this chapter, we present a computational model called the Proximity Cloud for bringing the cloud all the way to the edge to enable a new type of hybrid applications called edge apps in which application functionality is split between the edge and the data center cloud. We describe its implementation over OpenStack, as well as two functional edge apps we implemented and deployed in the Proximity Cloud prototype.

Numerous efforts have been made to cluster unused edge computing resources into a globally available compute pool, for example, BOINC [17], BOINC-based SETI [18], SEATTLE [26]. While the goal of these efforts is to harness compute cycles at the edge using a parallel computational workflow, the Proximity Cloud and edge app models are designed for general application services. In addition, the Proximity Cloud allows edge apps to execute in a coordinated manner in the edge as well as the data center. Such hybrid execution distinguishes the Proximity Cloud from other cloud computing models like Cloudlets [35] and Nanodatacenters [11], which focus solely on bringing the data center to the edge. Fog computing [20] presents a paradigm for a cloud extension called the "fog" to interface with IoT platforms like connected vehicles and smart grids. RACE [21] is a framework and system for specifying and executing a class of distributed real-time applications in the cloud based on real-time feeds collected from a large number of edge-devices. The Proximity Cloud is a more general model which can work with all types of edge services including IoT platforms and regular enterprise/home applications, and leverages a standard cloud platform rather than a specialized application level framework.

The Proximity Cloud is a natural extension to ISPs seeking to provide enhanced edge services to their customers. For example, ISPs like Comcast partition WiFi access on home WiFi access points/residential gateways into separate private and public wireless networks, with the public network acting as a WiFi hotspot to nearby Comcast customers [4]. While this is not full-blown Proximity Cloud functionality, it illustrates the ability of ISPs to manage black boxes at the edge.

The concept of the distributed Proximity Cloud raises several interesting open issues, e.g., application deployment strategies (i.e., where to place application instances), failure recovery (i.e., how to recover from failed edge nodes), distributed bandwidth resource management (i.e., how to share limited bandwidth resources at the edge), edge node security (i.e., how to protect edge app components hosted at the edge). The Proximity Cloud model defined in this chapter serves as a foundation to explore and research these various refinements.

References

1. Wikipedia.   http://en.wikipedia.org/wiki/Cold_boot_attack. Accessed: November 19, 2014.

2. Cisco. Cisco Nexus 1000V series switches. http://www. cisco.com/c/dam/en/us/solutions/collateral/switches/nexus-1000v-switch-vmware-vsphere/at_a_glance_c45-492852.pdf. Accessed: November 19, 2014.

3. Cloudera. Cloudera CDH. http://www.cloudera.com/hadoop/. Accessed: November 19, 2014.

4. M. Silbey, Comcast integrated private-public wifi, 2013. http://www.lightreading.com/mobile/carrier-wifi/comcast-turns-homes-into-hotspots/d/d-id/703027. Accessed: November 19, 2014.

5. Dummynet. http://info.iet.unipi.it/luigi/dummynet/. Accessed: November 19, 2014.

6. GStreamer. http://gstreamer.freedesktop.org. Accessed: November 19, 2014.

7. Hadoop Open Source Project. http://hadoop.apache.org/. Accessed: November 19, 2014.

8. Intel integrated performance primitives (Intel IPP). https://software.intel.com/en-us/intel-ipp. Accessed: November 19, 2014.

9. JPPF. http://www.jppf.org. Accessed: November 19, 2014.

10. Kinect for Xbox 360. http://www.xbox.com/kinect/. Accessed: November 19, 2014.

11. Nanodatacenters. http://www.nanodatacenters.eu. Accessed: November 19, 2014.

12. Net Index. http://www.netindex.com/. Accessed: November 19, 2014.

13. Nicira. http://nicira.com. Accessed: November 19, 2014.

14. Tinc VPN. http://www.tinc-vpn.org. Accessed: November 19, 2014.

15. Unlocking a LUKS encrypted root partition via SSH, 2008. http://www.debian-administration.org/articles/579. Accessed: November 19, 2014.

16. Cisco Systems Inc. Cisco global cloud index: Forecast and methodology, 2010-2015. White Paper, 2011.

17. D. P. Anderson. BOINC: A system for public-resource computing and storage. In *Fifth IEEE/ACM International Workshop on Grid Computing*, 2004.

18. D. P. Anderson, J. Cobb, E. Korpela, M. Lebofsky, and D. Werthimer. SETI@home: An experiment in public-resource computing. *Communications of the ACM*, 45, 56–61, 2002.

19. A. Basu. Intel AES-NI performance testing over full disk encryption. Technical report, Intel Corporation, Santa Clara, CA.

20. F. Bonomi et al. Fog computing and its role in the internet of things. In *ACM Sigcomm MCC*, Helsinki, Finland, 2012.

21. B. Chandramouli et al. RACE: Real-time applications over cloud-edge. In *ACM Sigmod*, Scottsdale, AZ, 2012.

22. J. Dean and S. Ghemawat. MapReduce: Simplified data processing on large clusters. *Commun. ACM*, 51, 107–113, 2008.

23. Docker. http://www.docker.com. Accessed: November 19, 2014.

24. OpenStack Folsom. http://www.openstack.org/software/folsom/. Accessed: November 19, 2014.

25. D. Hyde. A survey on the security of virtual machines. Technical report, Computer Science & Engineering Department, Washington University in St. Louis, MO, 2009.

26. C. Kim, M. Caesar, and J. Rexford. Floodless in SEATTLE: A scalable ethernet architecture for large enterprises. In *ACM Sigcomm*, Seattle, WA, 2008.

27. LXC. https://linuxcontainers.org/. Accessed: November 19, 2014.

28. NFV. http://en.wikipedia.org/wiki/Network_Functions_Virtualization. Accessed: November 19, 2014.

29. OpenStack Nova. http://nova.openstack.org. Accessed: November 19, 2014.

30. L. O'Gorman, Y. Yin, and T. Kam Ho. Motion feature filtering for event detection in crowded scenes. In *PRCA*, Tsukuba, Japan, 2012.

31. OnLive. http://games.onlive.com. Accessed: November 19, 2014.

32. Raspberry Pi. http://www.raspberrypi.org. Accessed: November 19, 2014.

33. B. Raghavan, K. Vishwanath, S. Ramabhadran, K. Yocum, and A. C. Snoeren. Cloud control with distributed rate limiting. In *Sigcomm*, Kyoto, Japan, 2007.

34. J. Rosenberg et al. STUN—Simple Traversal of User Datagram Protocol (UDP) through network address translators (NATs). RFC 3489, March 2003.

35. M. Satyanarayanan et al. The case for VM-based cloudlets in mobile computing. *IEEE Pervasive Computing*, 8(4), 14–23, 2009.

36. R. Tsuruta. The I/O bandwidth controller: dm-ioband performance report, 2008. http://lwn.net/Articles/270728/. Accessed: November 19, 2014.

37. J. Turnbull. *Hardening Linux*. Apress, New York, 2005.

38. C. Yang. Boosting OpenSSL AES encryption with Intel IPP. Technical report, Intel Corporation, Santa Clara, CA, 2012. Accessed: November 19, 2014.

39. D. Yun et al. Sub-meter accuracy 3D indoor positioning algorithm by matching feature points of 2D smartphone photo. In *Pacific PNT*, Honolulu, HI, April 2013.

CHAPTER 4

Virtual Mobile Networks in Clouds

Mohamad Kalil, Khalim Amjad Meerja, Ahmed Refaey,
and Abdallah Shami

CONTENTS

4.1 Introduction

The emerging era of new, upcoming data-oriented 5G wireless broadband networks have many new challenges to face. Foremost of them is that these networks should provide ubiquitous coverage in any given country. Ubiquitous coverage is providing unrelenting service in urban cities, rural country sides, and in between regions. This includes providing coverage over highways and railways that run through forest and hilly areas. Such kind of coverage requires considerable infrastructure investment by service providers (SPs). Furthermore, the number of users using wireless broadband services is growing tremendously each year to add burden to SPs [4,10]. This is primarily because of the limited availability of wireless radio spectrum resources. Second, wireless networks are becoming data-oriented, which is reducing the profit margins for the SPs. The capital expenditures (CAPEX) and operating expenditure (OPEX) are growing while expanding their network coverage and capacity [15].

It is not economically feasible for any single SP to have his presence in the entire country with his own dedicated infrastructure. So a potential feasible solution is that SPs can install their infrastructure resources in different regions of the country and share their resources with other SPs through a lease agreement. Since premises and infrastructure sharing brings down investment and operating costs substantially, this model is being looked into more aggressively. Also there are models where the limitedly available wireless radio spectrum is pooled and shared among SPs for an effective utilization of the scarce wireless radio spectrum resources [11,12,16,17].

A 5G wireless broadband network infrastructure will consist of an edge network called radio access network (RAN) located close to the coverage premises and the central packet switching core network (CN) connected with the Internet. The sharing of resources among the SPs may be at different levels starting at simple passive sharing of the physical sites, towers, power, cabling, etc. [13]. The next level of sharing (as we will see the details later) will be active sharing, where the RANs are shared by SPs. This may or may not include spectrum sharing between the SPs. Further, the sharing could also include sharing of both RANs and CNs between SPs. All these different combinations will lead to different scenarios, which will be discussed in this chapter in detail.

Most of the current 4G wireless network technologies use specialized network hardware, which is difficult to share among SPs effectively. Particularly for dynamic sharing of network resources among SPs, a completely new architecture is required for building the network. For this purpose, the concept of virtualization is introduced to realize most or all of the network device functionality in the software domain rather than implementing on specialized network hardware. The main advantage of this kind of approach is that the entire network can be built using commercial off-the-shelf (COTS) servers that are substantially cheaper compared to the specialized hardware. Since the network functionality is entirely in software domain, the capacity of the network can be increased by simply adding more COTS servers to the network. Under low-network demand, the servers can be freed gracefully to conserve energy resources. As a result, the cost of the infrastructure and its maintenance as well as the energy bill is reduced dramatically [2].

To reduce the investment and maintenance costs further, instead of purchasing COTS servers, SPs can simply purchase the computing capacity from the cloud. This scenario is particularly useful when the network demand is unpredictable. To avoid over provisioning or under provisioning the network capacity, the computing power can be purchased according to the actual required demand and pay only for the usage [2]. This is possible only due to virtualization of the wireless networks. Converting the 5G wireless broadband networks into cloud-based virtual wireless networks, simply called "virtual mobile networks in clouds," requires virtualizing both CN and RAN networks. While virtualization concepts from the wired networks can be directly applied to the CN networks, it is not straight forward to carry out RAN virtualization. The challenging aspects of RAN virtualization are user mobility, uncertain wireless channel conditions, and constant variation in the number of users on RAN network [12]. The RAN virtualization framework should take into account the varying channel capacity, the fading, and the interference. Furthermore, the wireless channel is a shared medium unlike point-to-point wired links. The high mobility of the users must be taken care of using seamless handover techniques. The quality of experience (QoE) is an important metric that must be maintained in performing virtualization in RAN.

This chapter describes RAN network virtualization in cloud in Section 4.2. The sharing of resources of virtualized 5G wireless broadband networks simultaneously by multiple mobile network operators (MNOs) or simply called SPs is discussed in Section 4.3. Then the system model for sharing of resources using wireless resource virtualization (WRV) framework is provided in Section 4.4. Binary integer programming (BIP) formulation of the resource sharing problem is discussed in Section 4.5. Then the approach to iteratively solve the BIP problem is presented in Section 4.6. Simulation results of the performance of the proposed BIP algorithm and the simplified iterative algorithm is presented and discussed in Section 4.7. Finally, the chapter ends with the summary in Section 4.8.

4.2   Virtualizing RAN Network in Cloud

Before delving into the actual sharing of wireless broadband networks among SPs, the framework for virtualization is introduced. It is important to understand the way by which virtualization is performed in order to understand network sharing. But, before this, the moving of RAN network services into the cloud is introduced.

4.2.1   Need for Cloud-Based RAN: C-RAN

The recent move for cloud-based RAN (C-RAN) architecture for future 5G wireless broadband networks is purely driven by the necessity and the need of the hour. The volume of users and their requirement for wireless mobile broadband services is increasing at an alarming pace. SPs have to address the problem of explosive growth in the network traffic, especially multimedia, through increasing the capacity of their wireless networks. Unfortunately, unlike the wired networks, it is not straightforward to expand the capacity of wireless broadband networks as they are limited by scarcity of spectrum resources and lesser available bandwidth. The bandwidth capacity is time varying and the number of users in these networks are constantly changing. In addition to that, the users are highly mobile and they need both coverage and high bandwidth links to the network. The wireless medium is a broadcast network leading to interference issues from other simultaneous transmissions. The resource block (RB) or media access control (MAC) scheduling algorithms are extremely complex, working on smaller time frame that requires enormous computing power as networks grow.

All these limitations must be addressed systematically and elegantly such that the proposed solution is both scalable and manageable. The first step in addressing all these issues is to keep the cell size small such as microcells, picocells, and user installed indoor femtocells. These small-sized cells are capable of providing higher bandwidths, which is necessary for delivering multimedia traffic to the wireless users. The frequencies of the spectrum can be reused over time and space. They are capable of accommodating more number of users per unit area. They are ideally suited for urban environments where the density of users is high and the users are less mobile. When the users start to move faster, they tend to increase the number of handovers among cells. On the other hand, large macrocells are ideally suited for addressing high mobility scenarios as they reduce the number of handovers. Macrocells are best suited for low density user areas like rural country sides. Maintaining all these cells of various sizes is extremely complex, time consuming, and requires large computation power. It is also expensive to maintain all these heterogeneous networks (HetNets).

It is now evident that there is a solution (for time being) to address the growing needs of the wireless services. However, new problems have emerged in terms of increasing complexity and requirement of large computation power. SPs can scale up their networks provided that they have the desired profit margins. However, the data-oriented nature of the 5G networks does not offer such huge profits as expected in return to the investments by SPs. Clearly, it is evident that innovative solutions are needed to keep the investment costs low while expanding the network capacity. C-RAN is the new architecture that is now looked into to provide a cost-effective solution to the investment problem faced by SPs.

4.2.2 Proposed C-RAN Architecture

Recently a new architecture for RAN is being looked in, which uses the services of cloud network. The main idea is to move the MAC and digital signal processing (DSP) to the cloud by separating eNB of RAN network into base station unit (BSU) and antenna unit (AU) [8]. The BSU will be located in the cloud and the AU will be present at the cell premises. This way the entire processing is moved to the cloud network. Locating the BSUs in cloud as a single pool will provide benefits of reducing the incurred costs (of both investment and maintenance), increasing the utilization efficiency and providing the platform that is flexible to new, rapid, and easy changes to the technology.

The entire baseband DSP and time-frequency RB scheduling is carried out by BSU located in the cloud. The AUs that are present at the cell sites will receive digitally sampled in-phase and quadrature phase (I/Q) sequence of bit streams from their respective BSUs located in the cloud. Interference is a major issue as SPs will be maintaining HetNets containing macrocells, microcells, picocells, and user installed home-based femtocells. As frequency planning is cumbersome when many users install femtocells indoors in an unplanned manner, a flexible approach to overcome intercell interference is needed. Cognitive radio (CR) based technology can be used by user installed femtocells and small picocells to avoid interference with the macrocells and microcells. The CR installed in AUs of femtocells and picocells use sensing mechanism that will sense the frequencies, which are being under usage by nearby cells to avoid interference.

CR technology used in AUs will sense spatiotemporally available frequencies around its premises. The sensing results will be transmitted to their BSUs located in the cloud, which are connected through the fronthaul network links. Upon receiving the feedback information on the available frequencies from remote AUs, the BSUs will further process this information to carry out frequency planning to completely avoid interference. The BSU will instruct the AUs which frequencies they have to use for their downlink and uplink communications with the mobile users. The AUs will then use these allocated frequencies to operate within their cell region. Since this mechanism uses both distributed feedback from AUs and centralized decision making at the BSUs in cloud, this frequency planning is called hybrid frequency planning. Such a hybrid frequency planning will not only reduce or eliminate mutual interference in HetNet environments consisting of macrocells, microcells, picocells, and femtocells, but it will also reduce the scheduling complexity at the BSUs in the cloud. Such a solution is highly scalable for the growing size of networks particularly in the high density urban areas like metro cities. Another benefit is that the cells can go offline during less traffic times and come online during high traffic times without causing much trouble for this hybrid scheduling that is carried out in the BSUs.

The architecture of BSU and AU is shown in Figure 4.1. It is shown that BSU is located in cloud and AU is located at the cell premises.

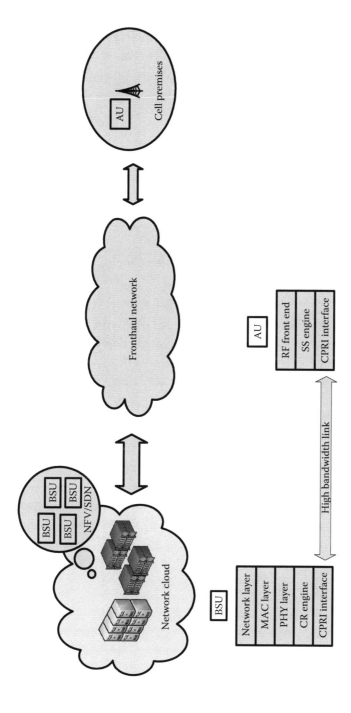

Figure 4.1 Proposed C-RAN architecture. BSU, base station unit; AU, antenna unit; MAC, media access control (link layer); PHY, physical (baseband digital signal processing); CR Engine, cognitive radio engine; SS Engine, spectrum sensing engine; CPRI, common public radio interface; NFV, network function virtualization; SDN, software defined networks. (From Meerja, K.A. et al., to appear in *IEEE Wireless Commun. Mag.*, 2014.)

They are connected through a fronthaul network with fiber optic links. The BSUs in the cloud and AUs at cell premises are placed such a way that the delay requirements are satisfied. Generally a BSU will manage more than one AU. The AU will mainly consist of transmitting and receiving antennas on the tower which transmit and receive radio signals from the mobile users. They additionally contain spectrum sensing (SS) units that are capable of sensing the frequencies that are in use and that are empty. For example, the AUs of the user installed femtocells can sense the availability of extremely high frequency (EHF) spectrum that is open for unlicensed transmission. This is around 60 GHz (V band) spectrum that is not used by either macrocells, microcells, or picocells. So, there will not be any interference between femtocells and other cells present in HetNet environment. This is very essential because femtocells are installed in an unplanned manner and by many indoor users, particularly in urban areas.

The main elements of AU are radio frequency (RF) frontend, SS unit and the common public radio interface (CPRI) unit. The RF frontend will further contain separate circuits that covert digital bit stream to analog signal through digital-to-analog (DAC) converter, and analog-to-digital (ADC) converter circuit to covert analog signal to bit stream. The CR-based SS engine in the AU will utilize the RF frontend for sensing the spatiotemporally available frequency spectrum in and around its cell premises. When idle frequencies are identified, the SS engine of the AU will transmit this information to BSU through CPRI interface. As entire intelligence is incorporated in BSU, AU will typically be a slave to BSU. Other than independent frequency sensing, AU will follow all the commands of BSU that maintains centralized control as desired by the MNOs.

The BSU, similar to eNB consists of network layer, MAC layer, and physical (PHY) layer. In addition to these, the cloud-based BSU contains CPRI unit and CR engine. CPRI unit in the cloud-based BSU receives bit stream that consists of control information regarding the availability of free channels based on AU sensing. The CR engine of the BSU processes this feedback control information regarding spectrum availability and sends them to the MAC layer. The MAC layer of the BSU makes scheduling decisions in a centralized manner based on the information received from its CR engine. The entire scheduling intelligence is located at the MAC layer that controls AUs and their transmission frequencies in both uplink and downlink directions. This hybrid scheduling mechanism is key to minimizing and eliminating mutual inter-cell and intra-cell interference in HetNet scenarios, which are common in urban cities.

4.2.3 C-RAN Virtualization

The main advantage of having a new split architecture for eNB in the form of BSU and AU and placing BSU into cloud is, we can pool all the BSUs that control AUs centrally from the cloud. Since the cloud has virtually unlimited processing power, the MNO can scale up and scale down the resource as and when needed based on the usage and online traffic demand. For this to happen, the BSUs in the pool are realized using virtualization software on COTS-based servers rather than implementing in specialized hardware. C-RAN virtualization makes the network flexible to the changes in the technology and gracefully adopting to new technologies with lesser glitches. During the transition phase both old and new technologies can coexist on virtual network elements created in software on COTS servers.

Network function virtualization (NFV) [6,7] framework can be used to virtualize the instances of BSUs. The cloud already offers various services in form of computing infrastructure as a service (IaaS), bare hardware as a service (HaaS), operating system platform as a service (PaaS), application software as a service (SaaS), and network of elements as a service (NaaS). Both computing infrastructure and network of elements as services (IaaS and NaaS) offered by the cloud network can be used in accordance with the NFV infrastructure (NFVI) framework to create BSU pool. As shown in Figure 4.2, NFVI is created over COTS hardware resources, which is controlled by a hypervisor. Hypervisor and hardware pool are maintained by the cloud network, through which, it offers compute power, storage, and networking capacity through IaaS and NaaS services. These services constitute the NFVI. SPs can use the NFVI services to create virtual network functions (VNFs), which in this case, is the realization of BSUs.

It is challenging to virtualize all the functionalities of a BSU. This is because BSUs host signal processing functions in their physical layer. For this reason, it is essential to initially carry out virtualization at higher layers, that is, at layers 2 and 3. Layer 3 of the BSU consists of control and data units that connect to the CN. CN can also be virtualized using NFV and SDN (software defined network) frameworks.

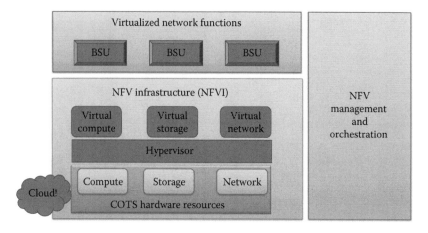

Figure 4.2   NFV framework for BSU pool in the cloud.

The SDN framework separates both control and data units of VNFs, which are also called virtualized network elements. The control unit is separated from VNFs and placed as a centralized control application on a SDN software server. Figure 4.3 shows a pictorial view of virtualized network elements as VNFs. All the key elements of the

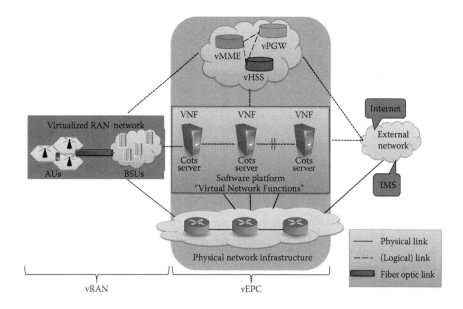

Figure 4.3   NFV/SDN framework for virtualized evolved packet core (vEPC) network.

evolved packet core (EPC) network are virtualized using NFV framework as VNFs such as vMME, vPGW, and vHSS–virtualized network functions for mobility management entity (MME), packet data network gateway (PGW), and home subscriber server (HSS). The layer 2 functionalities such as packet data convergence protocol, radio link control, and MAC can be realized as VNFs that can be centrally controlled by the SDN control unit.

4.2.4   Operation and Performance of C-RAN

In order to study the performance of C-RAN, the achievable data throughput in the proposed architecture for femtocells is presented [14]. The femtocells use CR technology for sensing the availability of the radio spectrum. The AU of the femtocell is capable of sensing simultaneously up to 16 frequency channels of 2 MHz bandwidth available in EHF radio spectrum. All those 16 freely available frequency channels are equally shared by the users in the femtocells. The MAC frame consists of eight equal length slots (T_{slot}). The length of T_{slot} is varied for throughput performance evaluation. The signal to noise ratio (SNR) is maintained at 10 decibels (dB). Also the sensing time $(T_{sensing})$ is varied to study the throughput performance. The femtocell AU and the users associated with the AU sense for a fraction of the slot length $(T_{sensing} < T_{slot})$.

For the MAC operation, the AUs of femtocells perform SS and send the result of sensing to their corresponding BSUs in the cloud. They identify channels that are not used by macrocell users, who are considered as primary users of the frequencies. Since EHF channels are not used by macrocell users, they are almost always idle for secondary transmissions by femtocell users. The central BSUs in the cloud will allocate the frequencies for data transfer to AUs based on their provided sensing feedback. The AUs will then allocate their assigned channels (up to 16 channels) to their associated users. Both femtocell AUs and their associated users will carry out sensing for $T_{sensing}$ time at the beginning of each slot on their assigned frequency channels. When the slot is identified as idle, they perform data transfer as secondary transmissions. When the slot is busy due to any primary user transmission, they withheld their transmission.

Whenever there is interference on the assigned channels or when these channels are no longer available for secondary transmission due to high usage activity by primary users, femtocell AUs will leave these channels and search for other set of frequencies that are not utilized around its premises. The secondary transmissions are carried out by femtocells on the newly negotiated channels with their BSUs in the cloud. The sensing mechanism will play a critical role on identifying the idle time slots in the MAC frame on any frequency channel. For any sensing mechanism such as simple stand-alone energy detection mechanism or advanced collaborative sensing mechanisms, the sensing time is critical. The higher is the sensing time, the better are the sensing results leading to accurate identification of the idle slots. Otherwise, an idle slot may be wrongly detected as busy that is called false-alarm. A wrong detection happens when a busy slot may be mistakenly detected as idle. Higher sensing time will keep false-alarm and wrong-detection probabilities to lower values, but at the cost of increased time overhead. The increased time overhead leads to lesser available time in the slot for actual data transmission, which results in lower data throughput. It is therefore necessary to select an optimal sensing time that will maximize the data throughput of the secondary transmissions.

To investigate the average achievable throughput for different number of users on the femtocell network, lengths of both T_{slot} and $T_{sensing}$ are varied as parameters. Initially, a smaller value for sensing time, of about 50 μs, is selected to study the throughput variation with increasing number of femtocell users under different slot lengths, as shown in Figure 4.4. When a single user is present in the femtocell, all the 16 channels acquired by the femtocell AU are assigned to that user, leading to higher throughput. When more number of users join the femtocell, those 16 channels are equally divided among the users that reduces the acquired throughput. Since it is very rare that there will be more than five users on the femtocell network, it can be seen from Figure 4.4 that considerable secondary user throughput can be achieved with the proposed MAC scheduling mechanism for C-RAN femtocell network. Figure 4.4 shows that throughput is higher when T_{slot} is 0.577 ms and improves slightly when T_{slot} is increased to 1 ms. When T_{slot} is increased further, the throughput decreases. This is because, when the sensing time is very small, there will be lot of sensing errors. Due to this, a lot of channel time is wasted due to wrong-detection and false-alarms. This is the main reason for the drop in the throughput with increasing slot length, when the sensing time is 50 μs.

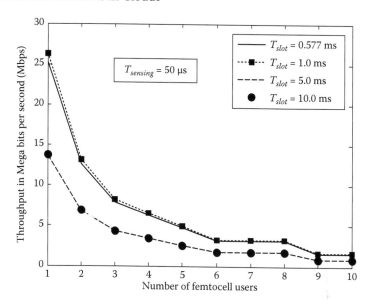

Figure 4.4 The data rate received by a user in femtocell under different slot lengths (T_{slot}). Sensing time, $T_{sensing}$ = 50 µs.

To see the improvement with increase in the sensing time, $T_{sensing}$ is now kept at 500 µs, and the throughput performance is reinvestigated over the same set of slot lengths as before. The throughput performance with 500 µs sensing time is shown in Figure 4.5. It can be seen that the new sensing time drastically reduces throughput of femtocell users when the slot length is 0.577 ms. This is because of excessive sensing overhead. Though more sensing time will improve sensing results, there is very little time left for actual data transmission as most of it is used for slot sensing. To better avail the improved performance of the sensing mechanism, the slot length has to be increased. The improvement can be readily seen in Figure 4.5, where it shows that the throughput increases with increase in slot length. It must however be noted that the throughput performance for 1.0 ms slot length in Figure 4.5 is less compared to that shown in Figure 4.4. This is because the sensing time overhead overweighs the improvement it offers when slot length is 1.0 ms. When the slot length is 5.0 ms or higher, the throughput performance shows improvement compared to the previous scenario. Figure 4.5 shows that it is possible to achieve 40 Mbps throughput when the slot length is 10.0 ms. It can be

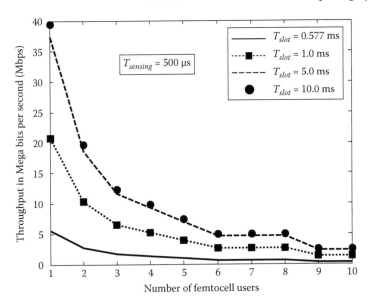

Figure 4.5   The data rate received by a user in femtocell under different slot lengths (T_{slot}). Sensing time, $T_{sensing} = 500\,\mu s$.

seen that over 10 Mbps user throughput can be achieved when four users are associated to the femtocell AU.

Now that the throughput performance of a femtocell as users vary was studied, the next important metric for study will be the variation of average throughput over the previous two important parameters, T_{slot} and $T_{sensing}$. The variation of average femtocell throughput under secondary transmission MAC scheme with increasing sensing time is shown in Figure 4.6. It can be seen that for small slot lengths such as 0.577 and 1.0 ms, the average throughput increases and then decreases with increasing sensing time of up to 750 μs. For 5.0 and 10.0 ms slot lengths, over 35 Mbps throughput is achieved when sensing time reaches 750 μs.

It can therefore be concluded that the selection of proper slot time and sensing time is crucial for achieving higher average throughput performance. When dynamics of sensing mechanism change with respect to its performance, the overall throughput performance will further vary. Better sensing mechanisms that require less sensing time will further improve the throughput performance of femtocell networks. With a simple and modest stand alone energy detection

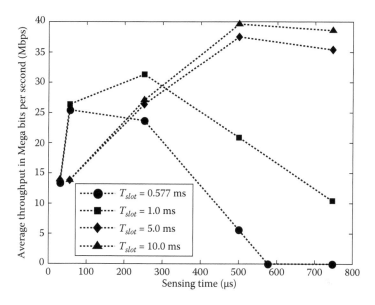

Figure 4.6   Variation of the average femtocell throughput with sensing time, under the absence of primary users, and for different slot lengths (T_{slot}). Sensing time, $T_{sensing}$ is varied from 25 to 750 µs.

sensing mechanism, a maximum average throughput performance of about 40 Mbps can be achieved over 16 channels, each of them having 2 MHz bandwidth, and under 10 dB SNR power.

4.3   Virtual Mobile Network Sharing Scenarios

Sharing of network infrastructure between MNOs was initially started with simple passive sharing of RAN premises such as physical sites, towers, power, cabling, and air conditioning units. Such passive sharing has tremendously brought down their investment, operation, and maintenance cost. With this in mind, MNOs started to look into further potential cost savings by including RAN active elements sharing. It is envisaged that MNOs can save up to 60 billion U.S. dollars worldwide, in 5 years period, and significant portion of these savings will be through active sharing of RAN elements. This has driven 3GPP standardization to identify five different active network sharing scenarios [5] as shown in Figure 4.7.

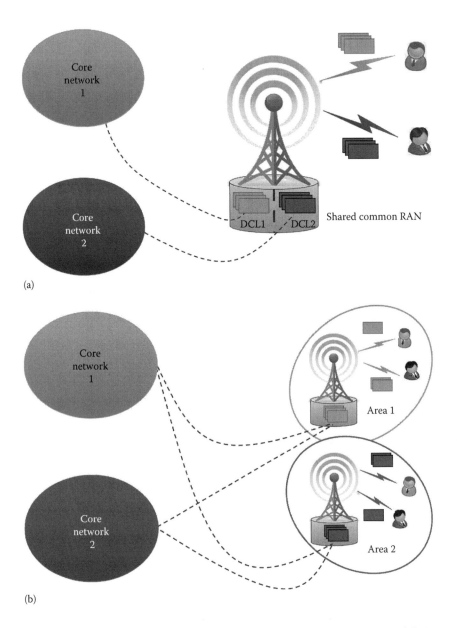

(a)

(b)

Figure 4.7 Scenarios for virtual mobile network sharing in 5G. (a) Scenario 1, (b) Scenario 2.   (*Continued*)

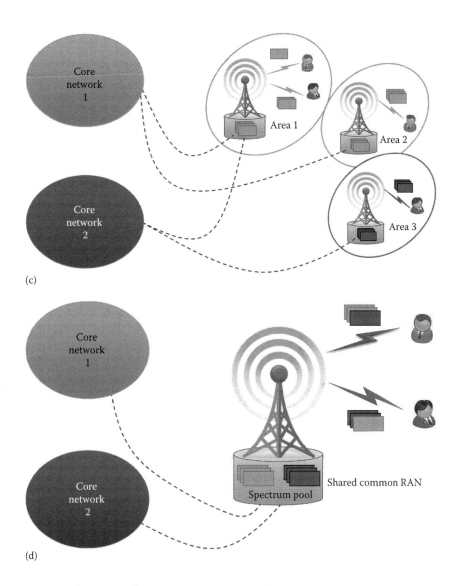

Figure 4.7 (*Continued*)   Scenarios for virtual mobile network sharing in 5G.
(c) Scenario 3, (d) Scenario 4.

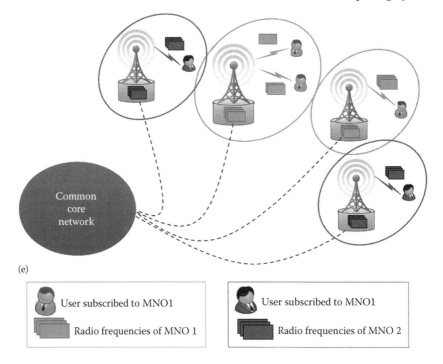

(e)

| User subscribed to MNO1 | User subscribed to MNO1 |
| Radio frequencies of MNO 1 | Radio frequencies of MNO 2 |

Figure 4.7 (*Continued*)   Scenarios for virtual mobile network sharing in 5G.
(e) Scenario 5.

4.3.1   Active Sharing: Scenario 1

The first scenario in the active RAN element sharing is shown in Figure 4.7a. The RAN elements which includes BSU and AU is shared between MNOs. In particular, the BSU pool in the cloud and the AUs at the cell premises that form a RAN network is shared by different MNOs. However, they have their own CNs. Their CNs are connected to the shared RAN network. Each MNO uses his own allocated radio frequencies to provide services to their subscribed users. As a result they connect directly to their dedicated carrier layer in a shared RAN. Figure 4.7a shows that two MNOs have their CNs connected to the shared RAN through their own dedicated carrier layer.

This type of sharing has the main advantage of effective utilization of BSU pool in the cloud. Also, MNOs need not have separate AUs installed in all the locations of their coverage. They can use a single AU in a given coverage area to provide their services. The MNOs can mutually own AUs in different locations and rent BSU pool of the cloud provider. It is also possible that there is a separate infrastructure provider who will be able to lease AUs in all the coverage regions from whom the MNOs can rent them. Thus by just owning the spectrum frequencies, they can provide services to the users as if they are running their own 5G network without actually owning the network infrastructure.

The only drawback with this kind of sharing is that MNOs are limited to provide their service in their allocated spectrum usage. Since the allocated spectrum to a MNO is only a portion of the already limited radio spectrum, expanding their service is difficult. Only a certain number of users will be served by each MNO with this kind of tight separation between the frequency spectrum used by different MNOs.

4.3.2   Active Sharing: Scenario 2

The second scenario of active RAN element sharing is shown in Figure 4.7b. In this scenario also, the MNOs have their own frequency spectrum to provide their services. They may use the common BSU pool provided by the cloud provider. Otherwise, some MNOs can own BSU pool and AUs in certain regions of the country. Other regions are covered by other MNOs. In regions where a particular MNO has no infrastructure laid for coverage, the infrastructure and spectrum resources of other MNO will be used to provide the service. Each MNO will have its own CN to provide 5G service.

Figure 4.7b describes this scenario where there are two MNOs providing service in the country. In Area 1, MNO 1 has the coverage with its laid RAN infrastructure. It uses its RAN infrastructure to serve its mobile broadband users. Since MNO 2 does not have RAN installed in Area 1, it connects its CN 2 to the RAN infrastructure of MNO 1 in Area 1 to provide service to its subscribed users. The service to the users of MNO 2 in Area 1 is provided over the frequencies allocated to MNO 1 over its carrier layer. The situation is vice versa in Area 2. In Area 2, MNO 2 has its RAN infrastructure. MNO 1 uses this infrastructure of MNO 2 to provide service to its subscribed users.

While this scenario gives the opportunity for a MNO to own the network in certain parts of the country and rely on the infrastructure of other MNOs in other parts of the country, the frequency spectrum utilization is very poor. For example in Figure 4.7b, in Area 1, the frequencies allocated to the MNO 2 were not used. To worsen the situation further, the frequency spectrum of MNO 1 in Area 1 is used by both MNOs. This will limit the expansion of services by both SPs.

4.3.3 Active Sharing: Scenario 3

The third scenario of active RAN sharing is to have RAN sharing similar to scenario 2 in only certain geographical areas. Outside these areas, coverage is provided by MNOs independently. This scenario is shown in Figure 4.7c. It shows that in Area 1 RAN sharing similar to scenario 2 is carried out. But Areas 2 and 3 are independently covered by these two MNOs.

This type of scenario is more practical for the transition purposes where sharing is not a forcible option. Those who want to own their network can own them and possibly share some parts of their network with other small MNOs. In this scenario, all MNOs have their own CN. Each MNO uses its own dedicated spectrum resources for providing services. The other drawback of this scenario is that some MNOs may not be able to cover all the regions of the country.

4.3.4 Active Sharing: Scenario 4

In this fourth scenario of active RAN elements sharing, frequency spectrum sharing is also included. Each of the different MNO owns a separate frequency spectrum and decides to pool their frequencies with others for common usage. This is described pictorially in Figure 4.7d in the case of having two MNOs. MNO 1 and 2 pool their allocated spectrum for common usage. Spectrum pooling will provide additional gain in the form of spectrum multiplexing gain. The BSU and AU can be rented from an infrastructure provider to provide services. This enables complete sharing of RAN resources including spectrum. However, MNOs have their own separate CNs.

4.3.5 Active Sharing: Scenario 5

The fifth scenario advocates sharing of CNs between MNOs. The RANs may or may not be shared and RAN sharing can follow any of the earlier scenarios. Sharing of CNs is shown in Figure 4.7e. Sharing of both CNs and RANs would be considered as the most aggressive network sharing scenario. With the advent of virtualization technology, it is very much feasible to achieve this level of sharing. For effective CN sharing, traditional virtualization techniques adopted in wired networks can be used. It is however challenging to share RANs efficiently because of additional challenges posed by wireless channels.

4.4   System Model for Sharing

The rest of the chapter takes scenario 4 (see Figure 4.8a) of the previous section as the system model for C-RAN sharing and it will be analyzed thoroughly under LTE flow assignment for mobile users. The entire RAN network including the spectrum is shared

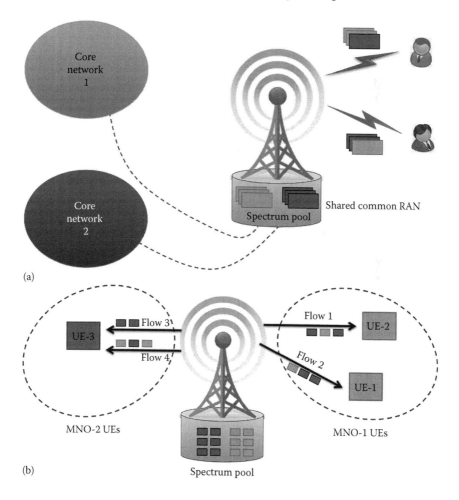

Figure 4.8   System model for virtual mobile network sharing. (a) Virtual RAN sharing: Scenario 4. (b) LTE flow assignment.

through pooling all the spectral resources of the MNOs. This system model also applies to scenario 5 of the previous section as one can assume that a single CN exists that connects to the shared RAN networks.

In this scenario two or more MNOs who have their own frequency spectrum join for sharing the cloud-based C-RAN by pooling their frequencies. The cloud-based BSU and AU which form the C-RAN network is shared among K different MNOs. The pool consists of N RBs that are shared by all MNOs over this C-RAN network. Each MNO may have its own CN, that is, they are associated to their own EPC. Otherwise, a single virtualized EPC (vEPC) network is shared by these K MNOs along with shared C-RAN network. The BSU will provide one or more radio flows to each mobile user depending on the number of connections established for multiple applications on that mobile device. However a single flow must be associated with only one user and cannot be shared in LTE. To ensure the quality of service (QoS), the assignment of flows is controlled by an admission control unit present in BSU. Under such a scenario, each MNO-k, in the set of MNOs belonging to $\mathcal{K} = \{1, 2, \ldots, K\}$, will have assigned F_k flows to its associated users. For example, Figure 4.8b shows that two MNOs share the cloud-based C-RAN network, where MNO-1 assigns 2 flows to its two users, that is, user 1 and user 2. On the other hand MNO-2 assigns 2 flows to its single user, user 3, which connects to two different applications. All through the analysis a downlink transmission scenario is considered.

The set of RBs $\mathcal{N}_k$ owned by each MNO-k is mutually exclusive so that

$$\mathcal{N}_p \cap \mathcal{N}_q = \emptyset, \quad \forall p, q \in \mathcal{K} \text{ and } \mathcal{N} = \bigcup_{k \in \mathcal{K}} \mathcal{N}_k. \tag{4.1}$$

The entire set of RBs $\mathcal{N}$ are available to all K MNOs as a pool of RF spectrum which will be shared according the negotiated agreement between them. The entire C-RAN network including spectrum pool is shared among the MNOs according to the set of agreements listed as follows:

1. Suppose that a flow f belonging to a particular MNO has a satisfaction level (SL) of Φ_f^{ns} in the absence of spectrum pooling.

That means, Φ_f^{ns} is the amount of SL received in case of no sharing of spectral resources among the MNOs on the C-RAN network. In case of no sharing, each MNO will use its own allocated spectral resource. The SL may be defined in terms of the amount of data throughput, worst case delay, tolerable jitter, or simply a combination of them. Then in case of spectrum pooling, any flow f should be able to experience SL of at least Φ_f^{ns}. This essential requirement will ensure that there is proper isolation between the MNOs on the shared C-RAN at the flow level. This condition will also protect SL of a particular flow from the traffic fluctuations on other flows belonging to other MNOs. Heavy traffic on MNOs are permitted when each of the flow of any MNO receives at least the SL when there is no sharing. Statistically, due to independent traffic fluctuations, each flow would receive more than Φ_f^{ns}, which is the maximum possible SL in case of no sharing. The resource sharing will provide extra gain in the SL due to multiplexing gain, which is referred as multi-MNO diversity gain (MDG).

2. Each MNO on the shared C-RAN will have different service level agreements with its users based on certain QoS or some sort of QoE negotiations during service provision. Further, each MNO has different operational requirements such as billing. Due to these reasons, each MNO will be using a different scheduling algorithm to satisfy its user demands based on quantitative QoS or qualitative QoE metrics, and billing policies. A particular MNO will be interested in providing fairness among its users and for this purpose, it uses proportional fair (PF) scheduling. Other MNOs may prioritize their revenue generation over user fairness and would simply use sum-rate maximization scheduling policy that allows users who pay more and who have better channel conditions to transmit more.

3. The MDG gain due to multi-MNO diversity will increase with the number of users in C-RAN. This is due to the fact that channel conditions are independent across users, which leads to increase in the aggregated channel capacity. The achieved MDG should be shared among all K MNOs present on the C-RAN according to the sharing agreements among them. By denoting

Φ_f^{ws} as the SL of flow f in the case of spectrum sharing through pooling, the MDG of this flow is given by

$$\Phi_f^{ws} - \Phi_f^{ns},$$

As a result, the total MDG achieved for MNO-k is given by

$$\Phi_k = \sum_{f=1}^{F_k} \Phi_f^{ns} - \Phi_f^{ws}. \tag{4.2}$$

MNOs can in general share MDG according to the following condition.

$$\bar{\Phi}_1 : \bar{\Phi}_2 : \cdots : \bar{\Phi}_K = \beta_1 : \beta_2 : \cdots : \beta_K \tag{4.3}$$

where

$\bar{\Phi}_1, \bar{\Phi}_2, ..., \bar{\Phi}_K$ are the average values of $\Phi_1, \Phi_2, ..., \Phi_K$, respectively

$\beta_k \in [0,1], \forall k$ are constants and satisfy

$$\sum_{i=1}^{K} \beta_k = 1. \tag{4.4}$$

4.4.1 Wireless Channel Model

The fading in the wireless channel between the AU and mobile user is assumed to be block Rayleigh fading. It is further assumed that the channel fading remains constant over the bandwidth of the RB assigned to a user. The channel varies independently over different RBs assigned to the users. Denoting the channel gain for any user u over a RB n as $G_{u,n}$, the received SNR for that user is given by

$$S_{u,n} = \frac{P \times G_{u,n}}{\sigma^2} \tag{4.5}$$

in Equation 4.5, σ^2 denotes the noise variance and P is the transmitted power of the AU along the downlink. In the system model, it is assumed that a constant power P is applied in transmission to all users across all RBs.

The duration of a subframe in LTE standard is 1 ms long and it contains 132 symbols. Therefore the size of a transport block that can actually be transmitted is found as [9]

$$T_r = \lfloor 132 \times \eta_r \rfloor \qquad\qquad (4.6)$$

where

η_r, is the spectral efficiency of modulation and coding scheme r with coding rate R

$\lfloor x \rfloor$ gives the largest integer number, which is less than or equal to u

The SNR that is received by the user determines the actual modulation and coding (MCS) scheme that should be used to send the transport block of data under a given block error rate (BLER). A simple way is to perform lookup operation onto the table that maps received SNR to the MCS scheme similar to the one given in Table 4.1. The mapping in Table 4.1 is for 10% BLER.

For illustration, if a user receives its downlink signal at 10 dB SNR on a particular RB, the MCS scheme that has to be used is indexed

Table 4.1 Modulation and Coding Scheme Used in LTE

Index (r)	Modulation	Coding Rate (R)	η_r	SNR (dB)
0	—	—	0 bits	> -6.7536
1	QPSK	78/1024	0.15237	$-6.7536 : -4.9620$
2	QPSK	120/1024	0.2344	$-4.9620 : -2.9601$
3	QPSK	193/1024	0.3770	$-2.9601 : -1.0135$
4	QPSK	308/1024	0.6016	$-1.0135 : +0.9638$
5	QPSK	449/1024	0.8770	$+0.9638 : +2.8801$
6	QPSK	602/1024	1.1758	$+2.8801 : +4.9185$
7	16QAM	378/1024	1.4766	$+4.9185 : +6.7005$
8	16QAM	490/1024	1.9141	$+6.7005 : +8.7198$
9	16QAM	616/1024	2.4063	$+8.7198 : +10.515$
10	64QAM	466/1024	2.7305	$+10.515 : +12.450$
11	64QAM	567/1024	3.3223	$+12.450 : +14.348$
12	64QAM	666/1024	3.9023	$+14.348 : +16.074$
13	64QAM	772/1024	4.5234	$+16.074 : +17.877$
14	64QAM	873/1024	5.1152	$+17.877 : +19.968$
15	64QAM	948/1024	5.5547	$> +19.968$

Source: 3GPP.TS.36.213 11.0.0, LTE; Evolved Universal Terrestrial Radio Access (EUTRA), Physical layer procedures, 2012.

as $r = 9$ in Table 4.1. The transport block size of this scheme is $T_r = 317$ bits with a BLER of 10%.

4.4.2 Utilization of C-RAN under Sharing

It would be useful to have a preliminary idea of what could be the utilization of a LTE standard C-RAN under spectrum sharing through a common pool. For this purpose the same downlink scenario is considered. Figure 4.9 shows the utilization of C-RAN under sharing with increase in the number of MNOs. For throughput evaluation, downlink transmissions are considered between AU of the C-RAN and the users of different MNOs. Exactly 20 users are considered for each MNO. A total of 100 RBs are pooled for C-RAN sharing. All the MNOs apply maximum throughput (MT) scheduler. The wireless channel between AU and each user is modeled as a block Rayleigh fading as discussed in previous section. The wireless channel is assumed to be constant over an RB bandwidth, but changes independently over RBs and users. The average SNR for every user is assumed to be 10 dB. An FTP traffic model is used to model the traffic for each user. Each user downloads one file, which has a size of 0.5 Mbytes. The time interval

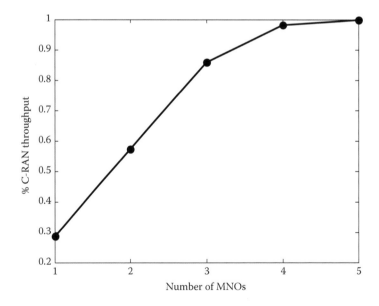

Figure 4.9 C-RAN throughput variation with increase in the number of MNOs.

between end of download of previous file and the user request for the next file is assumed to follow an exponential distribution with 5 s mean.

As mentioned earlier, Figure 4.9 shows the relationship between the number of MNOs and the relative throughput or % utilization achieved over the shared C-RAN. As number of MNOs increases, the C-RAN throughput increases as a result of two factors: (1) more MNOs improves MDG as a result of increasing the number of users in C-RAN; (2) increasing number of users reduces the probability of nonutilized RBs in cases such as the downloaded data is quite small and RBs are not used. However, saturation throughput performance is noticed when the number of MNOs is high. This is because C-RAN has a certain throughput threshold that cannot be exceeded.

4.5   Binary Integer Programming

In the absence of a C-RAN sharing, any MNO-k will be allocating a flow f to its user with SL of Φ_f^{ns} with its own set of $\mathcal{N}_k$ RBs. In case of C-RAN sharing, all of the $\mathcal{N}_k$ RBs belonging to K MNOs, where $k \in 1,2,...,K$, are pooled to form a common pool of $N = |\mathcal{N}|$ RBs, such that

$$\mathcal{N} = \bigcup_{i=1}^{K} \mathcal{N}_i \tag{4.7}$$

With pooling, it is expected that the SL for flow f will raise to Φ_f^{ws} due to MDG. However it is the resource allocation mechanism that is critical in realizing the full potential of achievable MDG due to multi-MNO diversity.

The most appealing scheme would be to use an appropriate utility function to carry out resource allocation on the shared C-RAN. Utility function is a means of quantitatively specifying the expected SL of a particular traffic type. In other words, utility function implements SL of a particular traffic type. Utility function varies with the type of traffic and nature of application. For instance, utility function of the best effort applications is realized by the achievable throughput. For real-time applications it is the worst case delay and tolerable jitter. For constant bit rate traffic, utility function is typically characterized by a

unit step function, which will be maximum if the observed throughput is greater than the required data rate and otherwise it is zero.

The objective of such utility-based resource allocation algorithm
would be to maximize the total utility factor of all flows belonging
to K MNOs. Each RB of the flow has a utility factor that matches the
traffic type that it is carrying. Denoting Z_f as the utility function of
the traffic type that is being carried by flow f, the utility factor of this
flow over RB n is given by $Z_{f,n}$. If $\mathcal{N}_f$ is the set of RBs that are assigned
to flow f, then the SL of this flow f, $\mathbf{Z}_f$, is given by

$$\mathbf{Z}_f = \sum_{n \in \mathcal{N}_f} Z_{f,n} \tag{4.8}$$

From this result, the objective of the resource allocation algorithm is
thus derived. It is required to maximize the SL of each and every flow
belonging to all K MNOs, spanning all N RBs. In mathematical notation, this is expressed as the following function which must be satisfied during each LTE subframe duration of 1 ms.

$$\max \sum_{n=1}^{N} \sum_{f=1}^{F} \mu_k Z_{f,n} \lambda_{f,n} \tag{4.9}$$

based on the condition that

$$\sum_{f=1}^{F} \lambda_{f,n} \leq 1, \ \forall n \in N \tag{4.10}$$

$$\Phi_f^{ws} \geq \Phi_f^{ns}, \ \forall f \tag{4.11}$$

$$\bar{\Phi}_1 : \bar{\Phi}_2 : \cdots : \bar{\Phi}_K = \beta_1 : \beta_2 : \cdots : \beta_K \tag{4.12}$$

where
  F is the total number of flows μ_k is a constant which is predefined
  for each MNO, and is used to fairly divide the MDG between all
  MNOs based on Equation 4.12
  $\lambda_{f,n}$ is a binary indicator defined as

$$\lambda_{f,n} = \begin{cases} 1, \text{ if the RB } n \text{ is assigned to flow } f \\ 0, \text{ otherwise} \end{cases}$$

The values for $\mu_k, \forall k$ will make sure that the constraint given by Equation 4.12 is always satisfied and they determine the MDG sharing

among all MNOs. The values for $\mu_k, \forall k$ can either be fixed, predetermined based on agreement between the MNOs or determined iteratively based on the SL observed in resource allocation process. There can be a situation where a particular MNO-k may choose to have higher MDG, that is, higher β_k, compared to other MNOs. In that case, it will increase it μ_k to a higher value by incrementing by a small positive constant δ.

4.5.1 MDG Sharing

Consider a C-RAN sharing by only two MNOs, namely MNO-1 and MNO-2. The setting of μ ensures to satisfy constraint in Equation 4.12, which determines how MDG should be distributed between the two MNOs. A simple controller is developed to achieve the required multi-MNO gain sharing. Let δ be a positive constant, the new value of μ incrementally increases or decreases as follows:

$$\mu_k[t] = \begin{cases} \mu[t-1]+\delta, & \text{if } \bar{\Phi}_k[t] < \bar{\Phi}_k \\ \mu[t-1]-\delta, & \text{if } \bar{\Phi}_k[t] > \bar{\Phi}_k \end{cases} \tag{4.13}$$

where $\Phi_k[t]$ is the average of Φ_k at t.

4.5.2 BIP Formulation

In this section, the WRV problem is formulated as a BIP problem. The problem in Equations 4.9 through 4.12 can be formulated in the general form of BIP optimization:

$$\mathbf{c}^T \mathbf{x} \tag{4.14}$$

subject to

$$\mathbf{A}\mathbf{x} \leq \mathbf{b} \tag{4.15}$$

vector $\mathbf{c}$ represents utility functions, and $\mathbf{x}$ is a binary decision vector maximizes Equation 4.9 and represents the term $\lambda_{f,n}$ in Equation 4.9. The constraints in Equations 4.10 and 4.11 are maintained by linear inequality constraints $\mathbf{A}\mathbf{x} \leq \mathbf{b}$ as follows:

1. Allocation constraints (Equation 4.10): Denote $\mathbf{I}_n$ as an identity matrix of size $n \times n$, and the matrix $\mathbf{A}_1$ as a matrix that contains all allocation constraints as follows.

$$\mathbf{A}_1 = \left[\underbrace{I_N, I_N, \ldots, I_N}_{F \text{ times}} \right]. \tag{4.16}$$

To enforce an exclusive RB assignment to only one flow, the vector $\mathbf{b}_1$ should be

$$\mathbf{b}_1 = \left[\underbrace{1, 1, \ldots, 1}_{N \text{ times}} \right]^T, \tag{4.17}$$

where $\mathbf{V}^T$ is the transpose of vector $\mathbf{V}$. The idea of this formulation can be best described with a small example. Suppose a system with the following parameter $N = 2$ and $F = 2$. The matrix $\mathbf{A}_1$ is

$$\mathbf{A}_1 = \begin{bmatrix} 1 & 0 & 1 & 0 \\ 0 & 1 & 0 & 1 \end{bmatrix} \tag{4.18}$$

and the vector $\mathbf{b}_1$ is

$$\mathbf{b}_1 = [1, 1]^T. \tag{4.19}$$

The vector $\mathbf{x}$ is defined as

$$\mathbf{x} = [\lambda_{1,1}, \lambda_{1,2}, \lambda_{2,1}, \lambda_{2,2}]^T,$$

The first row of Equations 4.18 and 4.19 matrices represents first RB assignment. The second row represents the second RB assignment. The inequality $\mathbf{A}_1 \mathbf{x} \leq \mathbf{b}_1$ ensures that $\lambda_{1,n} + \lambda_{2,n} \leq 1$, which forces a single allocation assignments for RB n between the two flows.

2. The constraints in Equation 4.11 can be formulated in the matrix $\mathbf{A}_2$ and the vector $\mathbf{b}_2$ as follows. Let $\mathbf{c}_f$ denote the vector that contains all the potential utilities to flow f as

$$\mathbf{c}_f = [Z_{f,1}, \cdots, Z_{f,N}]. \tag{4.20}$$

Define the matrix $\mathbf{A}_2$ as

$$\mathbf{A}_2 = - \begin{bmatrix} \mathbf{c}_1 & \cdots & \mathbf{0}_N \\ \vdots & \ddots & \vdots \\ \mathbf{0}_N & \cdots & \mathbf{c}_F \end{bmatrix}, \tag{4.21}$$

where $\mathbf{0}_N$ is a row vector of all zeros of size $1 \times N$. Vector $\mathbf{b}_2$ that satisfies the minimum LS constraints Equation 4.18 is

$$\mathbf{b}_2 = -[\Phi_1^{ns}, \dots, \Phi_F^{ns}]^T. \qquad (4.22)$$

The minus signs in Equations 4.21 and 4.22 reverses the direction of the inequality from $\Phi_f^{ws} \geq \Phi_f^{ns}$ to $-\Phi_f^{ws} \leq -\Phi_f^{ns}$ to be in the general form of BIP constraints.

Consequently, the linear inequality constraints that contain both the allocation and the minimum LS constrains are expressed as

$$\mathbf{A} = [\mathbf{A}_1, \mathbf{A}_2]^T \qquad (4.23)$$

and

$$\mathbf{b} = [\mathbf{b}_1, \mathbf{b}_2]^T \qquad (4.24)$$

4.6   Iterative Scheduling Algorithm

The resource allocation algorithm for assigning RBs to flows of MNOs in the shared C-RAN was formulated as BIP optimization problem in previous section. This optimization problem is NP-hard due to exponential increase in computational time with growing size of the problem. The size of the problem can easily grow due to increase in the number of MNOs on the shared C-RAN. To save computational power, it is essential to develop less computationally complex algorithms. For this purpose, an iterative algorithm is proposed to reduce the complexity in solving the BIP optimization problem that was previously discussed.

The iterative algorithm is presented in part at [11] and shown in Figure 4.10. This algorithm can be separated into two sequential blocks. The first part assigns RBs iteratively to the established flows of MNOs. During a single iteration, one RB is assigned to the flow, which has the highest difference between minimum SL Φ_f^{ns} and the current SL $\mathbf{Z}_f$. The RB assignment stops when all the flows receive SL that is greater than the minimum required, that is, until $\mathbf{Z}_f > \Phi_f^{ns}, \forall F$. The second part of the algorithm allocates the remaining RBs from the first part to the least satisfied MNO indicated. The MNO that is least satisfied has the largest difference between the agreed MDG $\bar{\Phi}_k$ and the experienced MDG given by $\Phi_k[t]$.

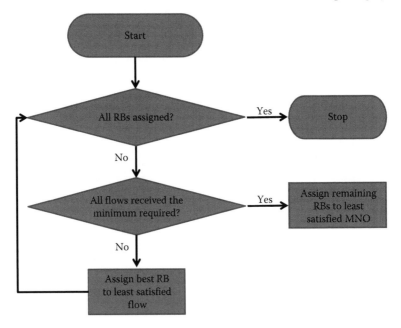

Figure 4.10 Iterative algorithm for RB allocation. (From Kalil, M. et al., *Wireless resources virtualization in LTE systems, IEEE INFOCOM 2014 Workshop on Mobile Cloud Computing*, May 2014.)

4.7   Numerical Results for Network Sharing

Now that original BIP optimization problem and the iterative algorithm to solve it have been discussed, the performance of these algorithms is presented in this section and they are also compared. The results discuss the performance of the system by taking two MNOs sharing the C-RAN system with spectrum pooling. The following two sections will present the parameters of evaluation and the results describing the performance of resource allocation algorithms of the shared C-RAN system.

4.7.1   Experimental Evaluation Parameters

Performance of original BIP and modified iterative resource allocation algorithms is considered in this section for the system in which the C-RAN is shared by two MNOs. Throughput is considered as the metric in the performance analysis. The following system parameters are assumed. First assumption is that the two MNOs apply different scheduling algorithms to allocate RBs to their flows. They allocate the RBs by taking them from the common shared pool. In particular, MNO-1 looks at the fairness and therefore carries out PF scheduling. The second MNO, MNO-2, cares more about the throughput performance and as a result applies MT scheduling.

PF scheduling of MNO-1 takes channel conditions into account while allocating the RBs to its flows. It provides fairness among users along with increasing the throughput performance. The utility factor of flow f in PF scheduling is defined by the following relation [3]:

$$Z_{f,n} = \frac{T_{f,n}}{\bar{T}_f} \tag{4.25}$$

where $T_{f,n}$ is the transport data block size that will be transmitted over RB n of flow f. This is computed from Equation 4.6. $\bar{T}_f$ is the historical average data transmitted by flow f. The PF scheduler will maximize the sum of all utility factors as follows:

$$\Phi^{PF} = \max \sum_{n \in \mathcal{N}_k} \sum_{f \in \mathcal{F}_k} \frac{T_{f,n}}{\bar{T}_f} \lambda_{f,n}$$
$$\text{such that } \sum_{f \in \mathcal{F}_k} \lambda_{f,n} = 1, \quad \forall n \in \mathcal{N}_k. \tag{4.26}$$

To maximize Φ^{PF}, the PF scheduler assigns RBs to users who have low historical average rates ($\bar{T}_f$), which gives them more chances of using the resources and thus imposes fairness between users, or to users who have good channel quality (high $T_{f,n}$) to maximize total throughput of the system. This will be achieved by having the following scheduling policy

$$S(f,n) = \arg\max_{f \in \mathcal{F}_k} \frac{T_{f,n}}{\bar{T}_f}, \tag{4.27}$$

where $S(f,n)$ represents the allocation of RB n to flow f.

In case of MT scheduling, the MNO will achieve MT by using the following utility factor [3]

$$Z_{f,n} = T_{f,n}. \tag{4.28}$$

and the maximization of the sum of utility factors is carried out using the following conditions.

$$\Phi^{MT} = \max \sum_{n \in \mathcal{N}_k} \sum_{f \in \mathcal{F}_k} T_{f,n} \lambda_{f,n}$$
$$\text{such that } \sum_{f \in \mathcal{F}_k} \lambda_{f,n} = 1, \quad \forall n \in \mathcal{N}_m. \tag{4.29}$$

Maximization of Φ^{MT} is achieved by the following scheduling policy.

$$S(f,n) = \arg\max_{f \in \mathcal{F}_k} T_{f,n}. \tag{4.30}$$

4.7.2 Experimental Evaluation Results and Discussion

The performance of the originally proposed BIP algorithm and the modified iterative algorithm is studied using simulations. The simulated scenario is shown in Figure 4.11. Two MNOs is assumed; MNO-1 owns five RBs, where MNO-2 own six RBs. As mentioned earlier, the C-RAN has a common RB pool, which is shared by two MNOs and each MNO has two users that are being served. Each MNO has two established flows assigned to its two users. One of the two users on each MNO is closer to the AU and receives an average SNR of 15 dB. The other user is far from the AU and receives an average SNR of 10 dB. All these users are assigned a single flow. The transmit power over every RB is 500 mW.

The first performance results are given in Figures 4.12 through 4.14, where a predefined values of μ_1, μ_2 are assumed for each experiment.

Figure 4.12 compares data rate of each flow in case of no sharing with the case of sharing for different μ_1 values using BIP formulation. in case of sharing the resources, each flow gains higher data rates than without sharing the resources. Users with same channel quality are treated differently. This is because the MNOs apply different scheduling policies. MNO-1 applies PF scheduling that gives users more

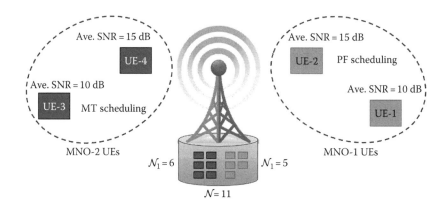

Figure 4.11 Simulated scenario.

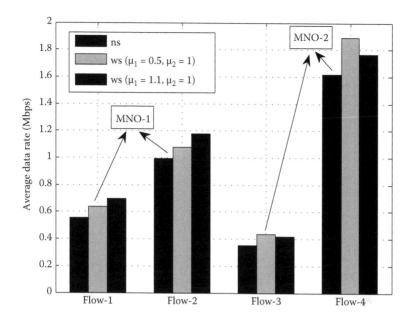

Figure 4.12   Flows' average data rate with no sharing and sharing cases for different values of μ_1 using BIP formulation.

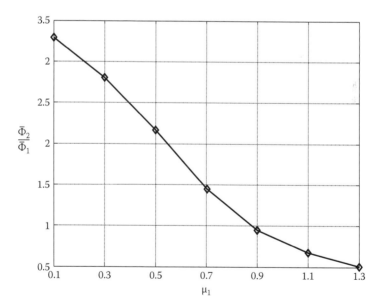

Figure 4.13   The relationship between μ_1 ($\mu_2 = 1$) and MDG shares between MNOs ($\bar{\Phi}_2/\bar{\Phi}_1$).

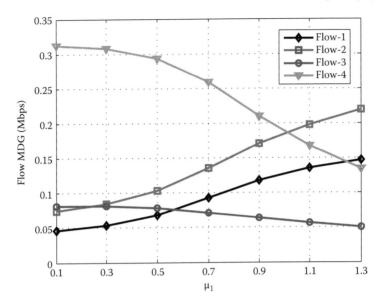

Figure 4.14 The relationship between μ_1 ($\mu_2 = 1$) and resource allocation efficiency represented by users throughput.

resources than MT scheduler in poor average channel conditions. As a result, the proposed BIP formulation allows MNOs to run custom flow scheduler on the same base station. Fixing μ_2 and increasing μ_1 bias the scheduling toward assigning more RBs to MNO-1. However, MNO-2's flows still receive data rates higher than in case of no sharing. Figure 4.13 illustrates the relationship between the parameters μ_1 and μ_2 and MDG of each MNOs. The value of μ_2 is fixed, while the value of μ_1 changes. As μ_1 increases the scheduler allocates more resource in favor of MNO-1's users which decreases $\bar{\Phi}_2/\bar{\Phi}_1$. Therefore, the constraints in Equation 4.12 can be satisfied by controlling the parameters μ_1 and μ_2. However, different settings of MDG result in different utilization of the resources as seen in Figure 4.14. As MT favors users in good channel conditions and utilizes the resource to the maximum, assigning more resources to MNO-2 (by lowering μ_1) results in more efficient utilization of the resources (higher aggregate cell throughput). Therefore, assigning more resources to MNO-2 (by lowering μ_1) results in more efficient utilization of the resources (higher aggregate cell throughput).

The second performance results are given in Figures 4.15 and 4.16. The values for μ_1 and μ_2 are found iteratively. The value of μ_2 is

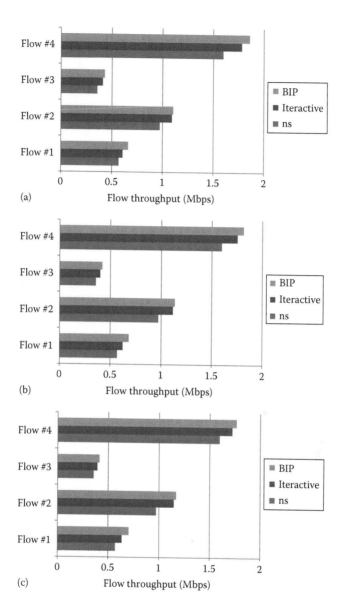

Figure 4.15 Performance evaluation of BIP and iterative approches. (a) Flow throughput for $\beta_1 = 0.4$ and $\beta_2 = 0.6$. (b) Flow throughput for $\beta_1 = 0.5$ and $\beta_2 = 0.5$. (c) Flow throughput for $\beta_1 = 0.6$ and $\beta_2 = 0.4$. (*Continued*)

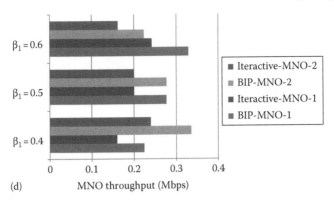

Figure 4.15 (*Continued*) Performance evaluation of BIP and iterative approches. (d) MDG per MNO. (From Kalil, M. et al., Wireless resources virtualization in LTE systems, *IEEE INFOCOM 2014 Workshop on Mobile Cloud Computing*, May 2014.)

set to 1 and μ_1 is varied until the constraints of Equation 4.12 are satisfied.

Figure 4.15 provides comparison between the original BIP resource allocation algorithm and the less computational modified iterative algorithm. It is worth pointing out that the running time of BIP solution is 67 times higher than that for the iterative algorithm. It can be clearly seen that sharing the common pool of RBs offers increased throughput performance. All the four flows belonging to the two MNOs experience throughput gain compared to the scenario when there was no sharing of RBs. The next thing that is observed is that the two users in both MNOs are treated differently based on observed channel conditions, which is indicated by the received SNR. The users that are closer to the AU have high SNR and therefore receive higher throughput when compared with the users that are far away from the AU. Since MNO-1 uses PF scheduler, the difference in throughput between its two users (flows 1 and 2) is less compared with the difference in throughput experienced by flows 3 and 4 who belong to MNO-2. This is because MNO-2 uses MT scheduling and as a result its closest user (flow 4) receives the highest throughput compared with all other three users on the C-RAN (both MNO-1 and MNO-2 combined). The flow that receives the least amount of throughput is flow 3 that also belongs to MNO-2.

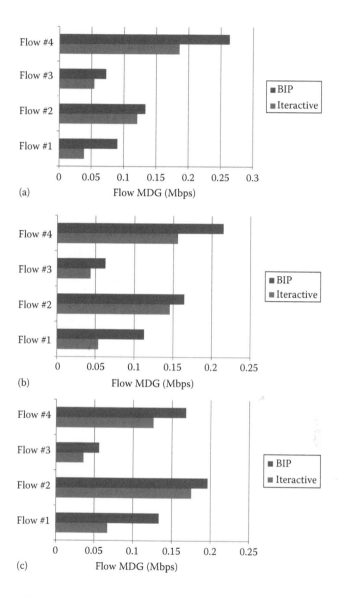

Figure 4.16   Performance evaluation of the BIP and the iterative approches.
(a) MDG distribution for $\beta_1 = 0.4$ and $\beta_2 = 0.6$. (b) MDG distribution
for $\beta_1 = 0.5$ and $\beta_2 = 0.5$. (c) MDG distribution for $\beta_1 = 0.6$ and
$\beta_2 = 0.4$.   (*Continued*)

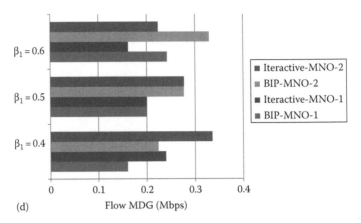

(d)

Figure 4.16 (*Continued*)   Performance evaluation of the BIP and the iterative approches. (d) Throughput per MNO. (From Kalil, M. et al., Wireless resources virtualization in LTE systems, *IEEE INFOCOM 2014 Workshop on Mobile Cloud Computing*, May 2014.)

The results in Figure 4.15 also shows the variation in throughput for all four users for different values of β_1 and β_2. It must be noted that sum of the values of β_1 and β_2 is equal to 1. To verify if MDG is proportionally distributed among the two MNOs according to β_1 and β_2, Figure 4.15d plots MDG values obtained for them in three scenarios with different settings for β_1 and β_2. When $\beta_1 = 0.4$ and $\beta_2 = 0.6$, MNO-2 receives greater MDG. When $\beta_1 = \beta_2 = 0.5$, both MNOs receive equal MDG. And when $\beta_1 = 0.6$ and $\beta_2 = 0.4$, MNO-1 receives higher MDG. Another conclusion that can be drawn from these results is that it is possible to achieve higher throughput and higher MDG when original BIP optimization algorithm is used. However is computationally intensive (67 times more) compared to the iterative algorithm that was developed to address this issue. But the difference of performance between these two algorithms is minimal. It can be seen that iterative algorithm is very handy to reduce the computational burden and at the same time achieve performance results that are close to the optimal values.

The variation of MDG, for four users belonging to MNO-1 and MNO-2, with respect to β_1 and β_2 is plotted in Figure 4.16. The performance of both original BIP algorithm and the modified iterative algorithm is presented. The results in Figure 4.16 show MDG for all four flows belonging to four users. MNO-1 shares its MDG between its flows 1 and 2 based on PF criteria. Whereas MNO-2 shares MDG for its flows 3 and 4 mainly based on higher achievable sum-rate criteria. As a result, MNO-1 achieves better fairness among its users on flows 1 and 2. In both MNOs, users that are close to AU, which receive higher SNR, are able to get higher MDG compared to the users that are far away from AU.

4.8 Summary

In this chapter, the concept of C-RAN virtualization is discussed. First of all, the virtualization of both RAN and CN to reduce CAPEX and OPEX for rolling out 5G networks is covered. The role of C-RAN in 5G networks for providing better services to the mobile users is presented. The performance of C-RAN that includes user installed femtocells is studied through simulations. The concept of actively sharing RAN elements and the vEPC is presented. The 3GPP standardization committee has proposed five active RAN sharing scenarios. Sharing of C-RAN that includes spectrum pooling under WRV framework is proposed. This new WRV incorporation in shared C-RAN system was modeled using BIP formulation. To reduce the computational complexity of BIP, an iterative algorithm was proposed to assign spectrum resources to the mobile users that belong to different MNOs. For both algorithms, their performance under C-RAN sharing was discussed and compared.

The proposed WRV framework allows MNOs to customize their schedulers independently while sharing the C-RAN. The framework also provides the required isolation between the MNOs to protect them from performance degradation due to C-RAN sharing. In particular this has been verified with the performance of two algorithms proposed for this kind of framework with a practical scenario of having two MNOs on shared C-RAN, each having two users with best and worst channel conditions. It is seen that the original BIP optimized resource allocation algorithm provides higher throughput and MDG. However, iterative algorithm is less complex and provides results that are close to optimal values. Both algorithms maintain the essential requirements WRV, that is, providing increased throughput and MDG performance through sharing and maintaining absolute isolation despite sharing spectrum resources.

References

1. 3GPP.TS.36.213 11.0.0. LTE; Evolved Universal Terrestrial Radio Access (E-UTRA); Physical layer procedures, 2012.

2. M. Abu Sharkh, M. Jammal, A. Shami, and A. Ouda. Resource allocation in a network-based cloud computing environment: Design challenges. *IEEE Communications Magazine*, 51(11):46–52, Nov. 2013.

3. F. Capozzi, G. Piro, L.A. Grieco, G. Boggia, and P. Camarda. Downlink packet scheduling in lte cellular networks: Key design issues and a survey. *IEEE Communications Surveys & Tutorials*, 15(2):678–700, May 2, 2013.

4. Cisco. Cisco visual networking index: Global mobile data traffic forecast update, 2013-2018. Technical report, 2013. http://www.cisco.com/c/en/us/solutions/collateral/service-provider/visual-networking-index-vni/white_paper_c11-520862.pdf.

5. X. Costa-Perez, J. Swetina, T. Guo, R. Mahindra, and S. Rangarajan. Radio access network virtualization for future mobile carrier networks. *IEEE Communications Magazine*, 51(7): 27–35, July 2013.

6. ETSI. Network function virtualization: An introduction, benefits, enablers, challenges, & call for action, white paper. 2012. http://portal.etsi.org/NFV/NFV_White_Paper.pdf.

7. H. Hawilo, A. Shami, M. Mirahmadi, and R. Asal. NFV: State of the art, challenges and implementation in next generation mobile networks (vEPC). *IEEE Network Magazine*, 28:18–26, 2014.

8. China Mobile Research Institute. C-RAN: The road towards green RAN, white paper, Version 2.5. Oct. 2011. http://labs.chinamobile.com/cran/wp-content/uploads/CRAN_white_paper_v2_5_EN.pdf.

9. M. Kalil, A. Shami, and A. Al-Dweik. QoS-aware power-efficient scheduler for LTE uplink. *IEEE Transactions on Mobile Computing*, 14:1672–1685, 2014.

10. M. Kalil, A. Shami, A. Al-Dweik, and S. Muhaidat. Low-complexity power-efficient schedulers for LTE uplink with delay-sensitive traffic. *IEEE Transactions on Vehicular Technology*, PP(99):1, 2014.

11. M. Kalil, A. Shami, and Y. Yinghua. Wireless resources virtualization in LTE systems. In *IEEE INFOCOM 2014 Workshop on Mobile Cloud Computing*, Toronto, Ontario, Canada, May 2014, pp. 363–368.

12. R. Kokku, R. Mahindra, H. Zhang, and S. Rangarajan. NVS: A substrate for virtualizing wireless resources in cellular networks. *IEEE/ACM Transactions on Networking*, 20(5):1333–1346, 2012.

13. D.-E. Meddour, T. Rasheed, and Y. Gourhant. On the role of infrastructure sharing for mobile network operators in emerging markets. *Computer Networks*, 55(7):1576–1591, 2011.

14. K. A. Meerja, A. Shami, and A. Refaey. Hailing the cloud empowered radio access networks (C-RAN). *IEEE Wireless Communications*, 22(1):122–129, 2015.

15. J.S. Panchal, R.D. Yates, and M.M. Buddhikot. Mobile network resource sharing options: Performance comparisons. *IEEE Transactions on Wireless Communication*, 12(9):4470–4482, Sep. 2013.

16. Y. Zaki, L. Zhao, C. Goerg, and A. Timm-Giel. LTE wireless virtualization and spectrum management. In *Wireless and Mobile Networking Conference (WMNC)*, Budapest, Hungary, 2010, pp. 1–6.

17. L. Zhao, M. Li, Y. Zaki, A. Timm-Giel, and C. Gorg. LTE virtualization: From theoretical gain to practical solution. In *Teletraffic Congress (ITC)*, San Francisco, CA, 2011, pp. 363–368.

CHAPTER 5

Task Admission Control for Cloud Server Pools

Haleh Khojasteh, Jelena Mišić, and Vojislav B. Mišić

CONTENTS

5.1   Introduction

Virtualization is an effective technique for maximizing the utilization of physical servers (hosts) in an Infrastructure-as-a-Service (IaaS) cloud data center, with a number of virtual machines (VMs) running on a given host [7]. Furthermore, energy expenditure of such servers can be minimized by pooling of hosts [8]. In this approach, servers are partitioned into a hot pool (always on and with VMs instantiated and ready to run), a warm pool (on but without VMs instantiated), and a cold pool with hosts turned off. Servers are, then, moved from one pool to another as needed to fulfill user requests or to conserve energy.

In this setup, maintaining desired performance levels is a major concern. In particular, accepting tasks when they arrive, only to reject them later, might lead to performance deterioration for tasks already taken into service and might also damage the reputation of the cloud

209

service provider. Admission control is an effective mechanism to enforce those performance levels and fulfill their respective service level agreements with users as required. In this chapter, we propose a simple yet effective admission control mechanism that can be easily added to an existing cloud center and investigate its performance as the function of system load and baseline partitioning of servers into pools.

This chapter is organized as follows: in Section 5.2, we briefly survey some related work in cloud admission control and resource allocation. Section 5.3 discusses the system model of cloud data center with the pooling mechanism and presents its performance. Section 5.4 describes the proposed admission control mechanism through two algorithms of increasing complexity and presents the performance improvements obtained in this way. In Section 5.5, a possible analytical model of the proposed admission control scheme is presented. Section 5.6 concludes the chapter and discusses some avenues for future work.

5.2   Related Work

Resource allocation and admission control in clouds have been the topic of much research effort in recent years. Some of the solutions were based on queueing theory: for example, [16] made use of the queueing information available in the system to make online control decisions. It used Lyapunov Optimization to design an online admission control, routing, and resource allocation algorithm for a virtualized data center. The algorithm considers a joint utility of the average application throughput and energy costs of the data center in decision making. In [10], authors proposed an autonomous scheme for admission control in cloud services aiming at preventing overloading, guaranteeing target response time, and dynamically adapting the admitted workload to compensate for changes in system capacity. They employed an adaptive feedback control scheme alongside with a queue model of the application. Cloud service, modeled using queuing theory, and controlled through adaptive proactive controllers that estimate whether services need some of the resources in the near future or not, was discussed in [1].

A Markov decision process (MDP) framework was proposed in [4] to model admission control in cloud, while approximate dynamic programming (ADP) paradigm was utilized to devise optimized admission policies. In [6], authors formulated an optimization problem for dynamic resource sharing of mobile users in Mobile Cloud Computing hot spot with a cloudlet as a semi-Markov decision process (SMDP) that was, then, transformed into a linear programming (LP) model to obtain an optimal solution. A unified framework of admission control and resource allocation was provided in [3] which modeled the systems dynamic behavior with a group of state-space models, scales between different desired operation points and used a set-theoretic control technique to solve admission control and resource allocation problems.

A technique for determining the effective bandwidth for aggregated flow was developed in [5] to make admission decisions using network calculus. Authors also examined the relationship between effective bandwidth and equivalent capacity for aggregated flow, while [13] modeled the admission control problem in a cloud using the general algebraic modeling system (GAMS) and solved it under provider-defined settings. Same authors in [9] presented a technique for admission control of a set of horizontally scalable services and their optimal placement into a federated Cloud environment. They have considered in their model that a request may also be partially accommodated in federated external providers, if needed or more convenient.

A session-based adaptive admission control approach for virtualized application servers was presented in [2]. Also, a session deferment mechanism was implemented to reduce the number of rejected sessions. In [12] three key characteristics of cloud services and IaaS management practices was identified which are burstiness in service workloads, fluctuations in VM resource usage over time and VMs being limited to pre-defined sizes only. Based on these characteristics, paper proposed scheduling and admission control algorithms that incorporate resource overbooking to improve utilization. A combination of modeling, monitoring, and prediction techniques was used to avoid exceeding the total infrastructure capacity.

The problem of optimal placement of VMs in clouds was tackled in [14] for minimizing latency. Complexity was reduced by recurring to a hierarchical split of the placement problem into two reduced complexity subproblems of choosing the data centers, then choosing the specific racks and servers, and applying a partitioning of the application graph to be placed. A resource allocation problem in [15] was formulated in which later tasks could reuse resources released by earlier tasks, and solved it with an approximation algorithm that can yield close to optimum solutions in a polynomial time.

5.3   System Model and Its Performance

5.3.1   System Model

The system model of a cloud server pool with a number of servers or hosts is shown in Figure 5.1. We assume that the system has a common input queue. Task requests arrive according to a Poisson process with arrival rate λ and they are served in the First In, First Out (FIFO) order. Without loss of generality, we assume that all servers are homogeneous as are the VMs; furthermore, we assume that a single VM image can satisfy all requests for a task.

To accommodate a request, the server selection module checks the server pools to find whether there is a server with sufficient number of idle VMs. The hot pool is checked first; if a server with sufficient number of idle VMs is found, the task can be serviced immediately. Otherwise, the warm pool is checked; if a warm server is to be used, it must first instantiate the required number of VMs which incurs some delay. Finally, the cold pool is checked, but bringing a server from the cold pool to the hot one requires additional delay for server start-up

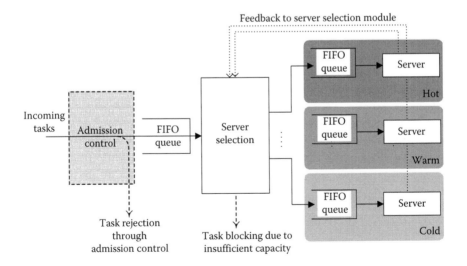

Figure 5.1   System model.

Table 5.1 Symbols and Corresponding Descriptions

Symbol	Description
N	Number of servers
λ	Incoming task rate
γ	Partitioning coefficient of hot, warm, and cold servers
ρ	Offered load
μ_{tot}	Overall service time
P_{blk}	Blocking probability
P_k	Probability of occurance of state k
λ_{org}	Full rate of incoming task rate
λ_{acd}	Accepted rate of task arrivals
d	Mean service time
T_1	Threshold state of full incoming task rate acceptance
T_2	Full occupied system state (the state which total number of VMs are used)
ρ_{av}	Average utilization of the system

and VM instantiation. Either way, the request is routed to the server FIFO queue where it awaits for the VM that will provision it.

If there is no server with sufficient capability, the request will be rejected. Requests can also be rejected if there is no space in the input queue.

Initial partitioning of available N servers is performed as follows: γN PMs are allocated to the hot and warm pools each, and the remaining $(1 - 2\gamma)N$ PMs are allocated to the cold pool. Servers are "upgraded" (i.e., moved from cold to warm, or from warm to hot pool) when requests need to be provisioned; they are "downgraded" as soon as requests finish service so that all VMs on a hot server become idle. To keep the system performance at an acceptable level, the number of hosts in the hot pool is never reduced below γN; the other two pools are allowed to change as needed.

Table 5.1 describes the symbols which have been used in this chapter.

5.3.2 Performance of the Original Pooled System

To investigate the performance of the pooled cloud system described earlier we have built a discrete event simulator using MATLAB® R2013a with the Simulink® component [11]. The results are shown in Figure 5.2. The diagrams in the top row show blocking probability as

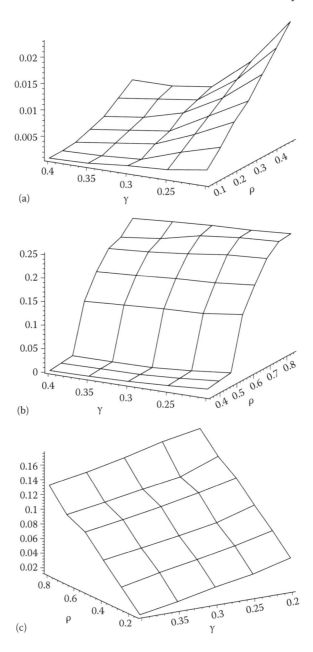

Figure 5.2   Task blocking probability and total delay. Blocking at service time 40 min, task arrival rate variable (a) from 100 to 600 per hour and (b) from 500 to 1200 per hour. (c) Blocking at task arrival rate 400 per hour, mean service time variable from 20 to 120 min.   (*Continued*)

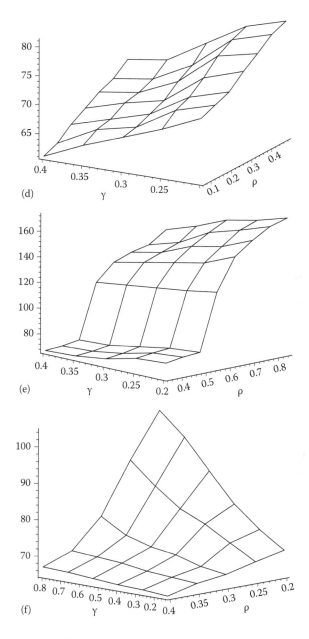

Figure 5.2 (*Continued*) Task blocking probability and total delay. Total delay (seconds) at service time 40 min, task arrival rate variable (d) from 100 to 600 per hour and (e) from 500 to 1200 per hour. (f) Total delay (seconds) at task arrival rate 400 per hour, mean service time variable from 20 to 120 min.

the function of the partitioning coefficient (i.e., default proportion of hot servers) γ and system load $\rho = \lambda/10N\mu_{tot}$, assuming the distribution of service times is exponential; diagrams in the bottom row show total delay under the same conditions. First two diagrams in each row are obtained by varying the task arrival rate under fixed service time; the rightmost diagram is obtained by varying the task service time under fixed task arrival rate. For clarity, we have separated the range of offered loads into two sub-ranges: one from $\rho = 0.1$ to 0.5, shown in the diagrams in the leftmost column, and another from $\rho = 0.4$ to 0.8, shown in the diagrams in the middle column; both correspond to mean service time of 40 min. The diagrams in the rightmost column were obtained under task arrival rate of 400 tasks per hour.

As can be seen, task blocking rate decreases with γ, which could be expected. When the task arrival rate increases, task blocking also increases. In the lower half of the range of load values, Figure 5.2a, this increase begins to show at low values of γ and higher values of ρ. However, when we look at the upper range of load values, Figure 5.2b, we see that the system actually enters saturation when task blocking rapidly increases beyond $\rho = 0.5$. The increase in blocking is more gradual when task arrival rate is fixed while mean service time increases, but the overall increase is quite steep, from around 0.02 (i.e., 2% of rejected tasks) to above 0.15 (i.e., 15%); again, this could be expected.

Saturation is also evident on the diagrams that show total task delay, especially in Figure 5.2e, even though the delay is only about twice of the maximum value in the lower part of the range, Figure 5.2d. Saturation is more noticeable in the diagram of delay obtained under fixed task arrival rate and variable service rate, Figure 5.2f.

Obviously, if we want to keep the cloud center out of saturation, in other words if we need to keep the delay and rejection rate low, some kind of admission control is desirable. By rejecting task arrivals that stand little to no chance of being serviced immediately, cloud service provider will improve the quality of service for the existing customers.

5.4   Simple Admission Control Mechanisms

The aim of applying admission control schemes to the pool management system is to keep the cloud system in the linear operation region with low blocking and reasonably small delay. We have seen that the partitioning coefficient γ affects blocking, and we can use this feature to improve performance. Therefore, admission control mechanism executes in two steps.

1. In the first tier, it adjusts the partitioning coefficient according to the mean task arrival rate. When the task arrival rate increases, the partitioning coefficient is increased to soften the impact of increased load, and vice versa. As task arrivals are random events, fluctuations of the task arrival rate are smoothed out by using an exponentially weighted moving average.

2. As the second tier of adjustment, if the probability of the task being rejected by the server selection module exceeds a predefined threshold P_{blk0}, the task is rejected outright. Again, the rejection probability is smoothed out via an exponentially weighted moving average.

In this manner, admission control filters out tasks so that most of the rejected ones are rejected outright, whereas only a small predefined percentage of them are rejected only after being processed by the server selection module, which takes additional time. This helps to maintain the rejection rate of the admitted tasks low. Furthermore, the elimination process of the admission control module can be used in a different way to improve performance, that is, to redirect superfluous task requests to another cloud server center instead; however, the elaboration of this approach is beyond the scope of the present work.

5.4.1 Performance of Admission Control

To investigate the performance of the admission control, we have defined two algorithms. The first one uses two distinct values of the partitioning coefficient, the default value γ_0 and the high-load value γ_1, as shown in Figure 5.3a; it is shown as Algorithm 5.1. The second one uses a "softer" changeover with three values for the partitioning coefficient: the default value γ_0, the mid-range value γ_1, and the high range value γ_2, as shown in Figure 5.3b; it is shown as Algorithm 5.2 here.

The resulting blocking probability under the two algorithms is shown in Figure 5.4, with initial partitioning coefficient and system load as independent variables. (Load is varied by varying the task

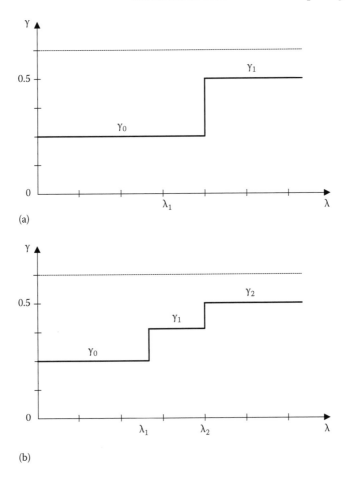

(a)

(b)

Figure 5.3   Dynamics of changing of γ according to admission control algorithms. (a) Changes of γ according to Algorithm 5.1 and (b) changes of γ according to Algorithm 5.2.

arrival rate.) For Algorithm 5.1, the thresholds were $\lambda_1 = 300$ tasks per hour, $\gamma_0 = 0.2$, and $\gamma_1 = 0.4$. In the blocking probability range of P_{blkl} to P_{blk0}, dropping rate (α) is a function of $\overline{\lambda}$, which means that by increasing of mean task arrival rate, dropping rate is going to increase as well. In defining α, we have been remotely motivated by packet dropping rate in random early detection mechanism used in transmission control protocol congestion control [17]. For Algorithm 5.2, the thresholds were $\lambda_1 = 300$ tasks per hour, $\lambda_2 = 400$ tasks per hour, $\gamma_0 = 0.2$, $\gamma_1 = 0.3$, and $\gamma_2 = 0.4$. Mean service time was

---

Algorithm 5.1 Admission control mechanism 1

Set $\lambda \longleftarrow \lambda_0$
while true do
    With each task arrival, recalculate $\bar{\lambda}$
    if $\bar{\lambda} > \lambda_1$ **then**
        Set $\gamma \longleftarrow \gamma_1$;
    else
        Set $\gamma \longleftarrow \gamma_0$;
    end if
    With each task rejection, recalculate $\overline{P_{blk}}$
    if $P_{blkl} < \overline{P_{blk}} < P_{blk0}$ **then**
        Drop most recently arrived task with the rate of $\alpha = f(\bar{\lambda})$;
    else if $\overline{P_{blk}} \geq P_{blk0}$ **then**
        Drop most recently arrived task;
    end if
end while

Algorithm 5.2 Admission control mechanism 2

Set $\lambda \longleftarrow \lambda_0$
while true do
    With each task arrival, recalculate $\bar{\lambda}$
    if $\bar{\lambda} > \lambda_2$ **then**
        Set $\gamma \longleftarrow \gamma_2$;
    else
        if $\bar{\lambda} > \lambda_1$ **then**
            Set $\gamma \longleftarrow \gamma_1$;
        else
            Set $\gamma \longleftarrow \gamma_0$;
        end if
    end if
    With each task rejection, recalculate $\overline{P_{blk}}$
    if $\overline{P_{blk}} \geq P_{blk0}$ **then**
        Drop most recently arrived task;
    end if
end while

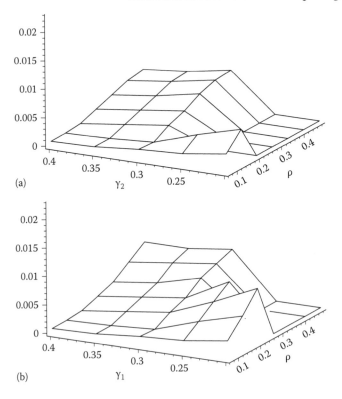

Figure 5.4   Task blocking probability with admission control according to (a) Algorithm 5.1 and (b) Algorithm 5.2.

kept at 40 min. As can be seen, both algorithms, together with pool management of the proposed approach, manage to maintain the overall blocking probability within reasonably low range. Interestingly enough, the "softer" adjustment provided by Algorithm 5.2 is less successful in keeping the overall blocking probability low than the "harder" one provided by Algorithm 5.1.

As further validation of the efficiency of Algorithm 5.1, we have plotted the task blocking rate of the system without admission control, with the task arrival rate periodically increased from 100 to 700 tasks per hour after every 250 s. Mean task service time was fixed at 40 min. As can be seen from the timing diagram in Figure 5.5b, task blocking rate is steadily increasing with each increment of the task arrival rate. However, when admission control according to Algorithm 5.1 is applied, task blocking rate is kept reasonably

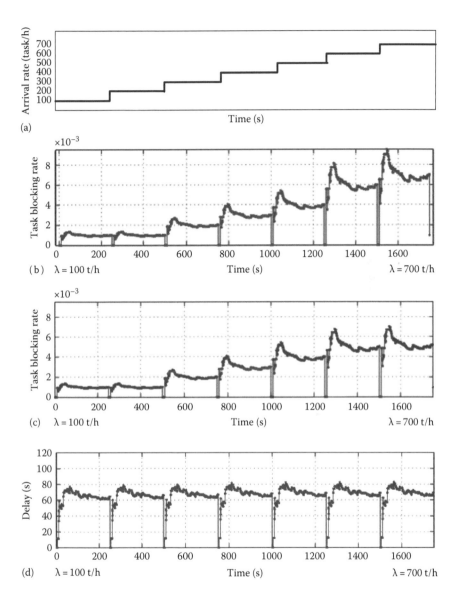

Figure 5.5   Test case of task arrival rate variable from 100 to 700 per hour, service time 40 min. (a) Task arrival rate changes during the test time. (b) Task blocking rate without admission control. (c) Task blocking rate with admission control (Algorithm 5.1). (d) Total delay with admission control (Algorithm 5.1).

constant, as shown in Figure 5.5c. The initial increase is due to the smoothing algorithm, which takes some time after the change in task arrival rate to adjust the mean task arrival rate as well as the mean blocking rate. Mean delay that tasks experience is also kept reasonably constant, as shown in Figure 5.5d, with the same caveat about transitory regime as above. In fact, the delay even decreases slightly, which is due to the fact that some tasks are not admitted in the first place, and therefore do not affect the performance.

In another test case which its results have been presented in Figure 5.6, we have increased the task arrival rate from 100 to 700 tasks per hour and then we have decreased the rate to 100 tasks per hour periodically after every 250 s. Mean service time was set to 40 min. It can be seen that in Figure 5.6c, using Algorithm 5.1 is improving the task blocking rate. Moreover, Figure 5.6d and e presents the changes of task rejection rate and γ respectively; these

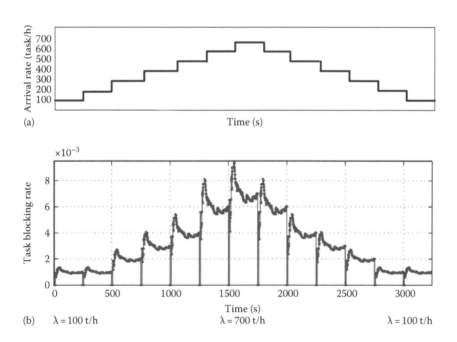

Figure 5.6 Test case of task arrival rate variable from 100 to 700 per hour and down to 100 per hour, service time 40 min. (a) Task arrival rate changes during the test time. (b) Task blocking rate without admission control.   (*Continued*)

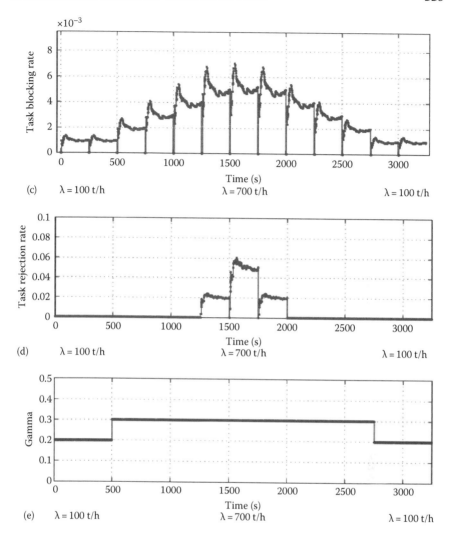

Figure 5.6 (*Continued*)   Test case of task arrival rate variable from 100 to 700 per hour and down to 100 per hour, service time 40 min. (c) Task blocking rate with admission control (Algorithm 5.1). (d) Task rejection rate with admission control (Algorithm 5.1). (e) γ change with admission control (Algorithm 5.1).

are the controlling parameters in Algorithm 5.1 and when task arrival rate increases, task rejection rate and γ will increase as well to reduce the task blocking rate. On the other hand, with decreasing of task arrival rate repeatedly, γ will be reduced to its minimum value and task rejection rate will get to zero.

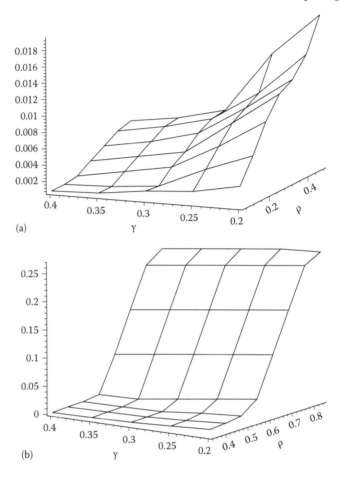

Figure 5.7   Task blocking probability with admission control (Algorithm 5.1 without considering task blocking threshold of P_{blk0}). Blocking at service time 40 min, task arrival rate variable (a) from 100 to 600 per hour and (b) from 500 to 1200 per hour.

In another test case presented in Figure 5.7, we have increased the task arrival rate from 100 to 1200 tasks per hour and in a similar case in Figure 5.8, we have increased the task arrival rate from 100 to 1000 tasks per hour periodically after every 250 s and mean service time has been fixed to 40 min in both figures. In this test case, we have applied Algorithm 5.1 to control the admission of arriving tasks. Although we have ignored the task blocking threshold of P_{blk0} to get into the higher rates of incoming tasks and examine the task blocking rate. Meanwhile, the task rejection rate is increased according to the increasing of incoming task rate. As can be seen, in Figure 5.7a, task blocking rate is less than the case of not using any admission control mechanism (Figure 5.2a). Furthermore, when task arrival rate increases in the ranges of 500–1200 tasks per hour (Figure 5.7b), task blocking rate will get into the saturation regime in higher rates of task arrival compared to Figure 5.2b which no admission control has been applied. In the timing diagrams in Figure 5.8b and e, task

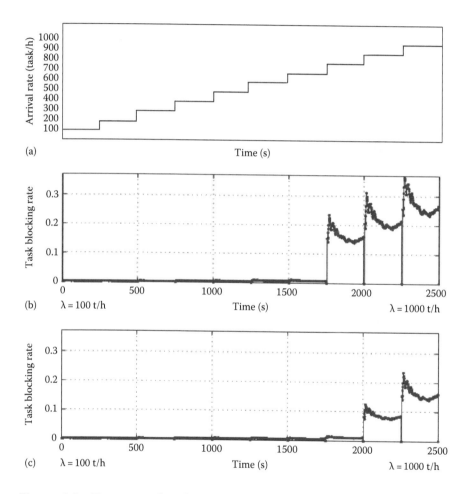

Figure 5.8 Test case of task arrival rate variable from 100 to 1000 per hour, service time 40 min. (a) Task arrival rate changes during the test time. (b) Task blocking rate without admission control. (c) Task blocking rate with admission control (Algorithm 5.1 without considering task blocking threshold of P_{blk0}). (*Continued*)

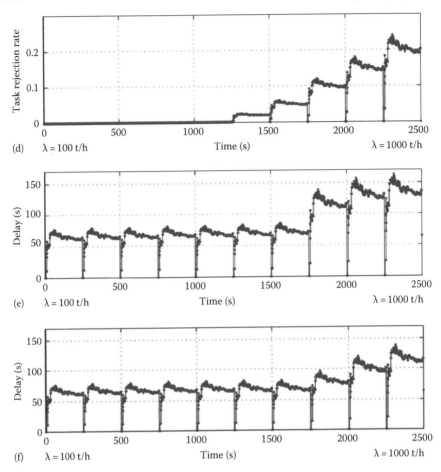

Figure 5.8 (*Continued*)   Test case of task arrival rate variable from 100 to 1000 per hour, service time 40 min. (d) Task rejection rate with admission control (Algorithm 5.1 without considering task blocking threshold of P_{blk0}). (e) Total delay without admission control. (f) Total delay with admission control (Algorithm 5.1).

blocking rate and total delay are increasing respectively with each increment of the task arrival rate. When Algorithm 5.1 is applied to model, task blocking rate and delay have been reduced as shown in Figure 5.8c and f. According to Figure 5.8d as the incoming task rate increases, the applied admission control mechanism will increase the task rejection rate to keep the total delay and task blocking below the threshold.

5.5 Analytical Model of Admission Control

Admission control scheme can be modeled as a Markov model; as an example, consider the chain shown in Figure 5.9. In this setup, T_1 and T_2 represent the thresholds of full task acceptance and full task rejection, respectively. Namely, when the mean blocking probability of the system is below the first threshold, admission control accepts all incoming tasks. As the result, T_1 is the last state where all of the arrived tasks will be admitted; however, some of them may be rejected later by the server selection module. Beyond this threshold, admission control begins to drop some of the incoming tasks. T_2 is the full rejection threshold state beyond which all incoming tasks will be rejected.

Probabilities of going from state i to state $i+1$ (i.e., that the task is accepted by the admission control) and the accepted rate of incoming tasks, λ_{adc}, can be set as

$$
\lambda_{adc} = \begin{cases} \lambda_{org}, & \text{state } i = 0 - T_1 \\ \lambda_{org}(1 - \frac{i-T_1}{T_2-T_1}), & \text{state } i = T_1 + 1 - T_2 \\ 0, & \text{state } i > T_2 \end{cases} \tag{5.1}
$$

where λ_{org} is the incoming task arrival rate. Then, the state probabilities of the Markov model can be calculated as

$$
P_k = \begin{cases} \dfrac{1}{k!}\left(\dfrac{\lambda_{org}}{d}\right)^k \dfrac{1}{DD} & 0 \leq k \leq T_1 \\[3ex] \dfrac{\lambda_{org}^{T_1} \prod_{i=1}^{k-T_1} \lambda_i}{k!\, d^k} \dfrac{1}{DD} & T_1 < k \leq T_2 \end{cases} \tag{5.2}
$$

where

$$
DD = \sum_0^{T_1} \frac{1}{i!}\left(\frac{\lambda_{org}}{d}\right)^i + \left(\frac{\lambda_{org}}{d}\right)^{T_1} \sum_{T_1+1}^{T_2} \frac{\prod_{i=1}^{i-T_1} \lambda_i}{i!\, d^{i-T_1}} \tag{5.3}
$$

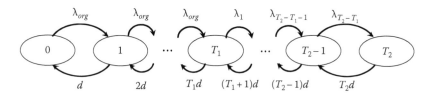

Figure 5.9   Markov chain for admission control according to Algorithm 5.2.

and d is the mean service time. The average utilization of the system ρ_{av} and the blocking probability of T_1 threshold P_{blk1} is given by

$$\rho_{av} = \frac{1}{T_2} \sum_{k=0}^{T_2} k P_k \qquad (5.4)$$

$$P_{blk1} = \sum_{k=T_1}^{T_2} P_k \qquad (5.5)$$

The last two equations are considered as controlling equations of the system; in case of choosing appropriate values for these two controlling parameters, the equations can be solved to obtain T_1 threshold.

5.6 Conclusion

In this chapter, we have examined the behavior of a cloud center in different operating areas where servers are pooled into hot, warm, and cold groups. Also, two algorithms for task arrival admission control are presented to keep the cloud system in the nonsaturation operating area. These admission control algorithms execute in two steps: first step is adjusting the partitioning coefficient according to the mean task arrival rate; the second tier is tracking the task blocking probability in the system and adjusting the task acceptance rate according to the predefined task blocking probability thresholds. Our simulation results confirm that the proposed admission control algorithms improve system performance. However, adjustments provided by Algorithm 5.1 appeared to be more successful in keeping the overall blocking probability low than the adjustments provided by Algorithm 5.2. Although, the cost of this success is dropping some of the incoming tasks in the linear to transition operating regions in Algorithm 5.1.

In our future work, we will provide multiple classes of tasks in the server pool management model that have different requirements and priorities, and we will adjust our solution according to these new requirements.

References

1. A. Ali-Eldin, J. Tordsson, and E. Elmroth. An adaptive hybrid elasticity controller for cloud infrastructures. In *IEEE Network Operations and Management Symposium (NOMS)*, Maui, HI, pp. 204–212, April 2012.

2. A. Ashraf, B. Byholm, and I. Porres. A session-based adaptive admission control approach for virtualized application servers. In *Fifth IEEE International Conference on Utility and Cloud Computing (UCC)*, Chicago, IL, pp. 65–72, November 2012.

3. D. Dechouniotis, N. Leontiou, N. Athanasopoulos, G. Bitsoris, and S. Denazis. ACRA: A unified admission control and resource allocation framework for virtualized environments. In *Eighth International Conference on Network and Service Management (CNSM) and Workshop on Systems Virtualiztion Management (SVM)*, Las Vegas, NV, pp. 145–149, October 2012.

4. Z. Feldman, M. Masin, A. N. Tantawi, D. Arroyo, and M. Steinder. Using approximate dynamic programming to optimize admission control in cloud computing environment. In *Winter Simulation Conference 2011 (WSC)*, Phoenix, AZ, pp. 3153–3164, December 2011.

5. Y. He, J. Huang, Q. Duan, Z. Xiong, J. Lv, and Y. Liu. A novel admission control model in cloud computing. Januaury 2014. Available from http://arxiv.org/abs/1401.4716v2. Last visited: September 1, 2014.

6. D. T. Hoang, D. Niyato, and P. Wang. Optimal admission control policy for mobile cloud computing hotspot with cloudlet. In *IEEE Wireless Communications and Networking Conference (WCNC)*, Shanghai, China, pp. 3145–3149, April 2012.

7. H. Khazaei, J. Mišić, and V. B. Mišić. Performance of cloud centers with high degree of virtualization under batch task arrivals. *IEEE Transactions on Parallel and Distributed Systems* 24(12):2429–2438, December 2013.

8. H. Khojasteh, J. Mišić, and V. B. Mišić. Characterizing energy consumption of IaaS clouds in non-saturated operation. In *Proceedings of the IEEE INFOCOM 2014 Workshop on Mobile Cloud Computing*, Toronto, ON, Canada, pp. 398–403, April 2014.

9. K. Konstanteli, T. Cucinotta, K. Psychas, and T. Varvarigou. Elastic admission control for federated cloud services. In *IEEE Transactions on Cloud Computing*, 2(3):348–361, July 2014.

10. N. Leontiou, D. Dechouniotis, and S. Denazis. Adaptive admission control of distributed cloud services. In *International Conference on Network and Service Management (CNSM)*, Niagara Falls, ON, Canada, pp. 318–321, October 2010.

11. The MathWorks Inc. Simulink simulation and model-based design. Website. http://www.mathworks.com/products/simulink/. Last accessed: September 1, 2014.

12. L. Tomas and J. Tordsson. Improving cloud infrastructure utilization through overbooking. In *ACM Cloud and Autonomic Computing Conference (CAC)*, Miami, FL, pp. 1–10, August 2013.

13. K. Konstanteli, T. Cucinotta, K. Psychas, and T. Varvarigou. Admission control for elastic cloud services. In *Fifth IEEE International Conference on Cloud Computing (CLOUD)*, Honolulu, HI, pp. 41–48, June 2012.

14. M. Alicherry and T. V. Lakshman. Network aware resource allocation in distributed clouds. In *IEEE INFOCOM*, Orlando, FL, pp. 963–971, March 2012.

15. F. Chang, J. Ren, and R. Viswanathan. Optimal resource allocation in clouds. In *Third IEEE International Conference on Cloud Computing (CLOUD)*, Miami, FL, pp. 418–425, July 2010.

16. R. Urgaonkar, U. C. Kozat, K. Igarashi, and M. J. Neely. Dynamic resource allocation and power management in virtualized data centers. In *IEEE Network Operations and Management Symposium (NOMS)*, Osaka, Japan, pp. 479–486, April 2010.

17. S. Floyd and V. Jacobson. Random early detection (RED) gateways for congestion avoidance. *IEEE/ACM Transactions on Networking* 1(4):397–413, August 1993.

CHAPTER 6

Modeling and Analysis of Single-Hop Mobile Cloudlet

Yujin Li and Wenye Wang

CONTENTS

A *mobile cloudlet* is a set of resource-rich mobile devices—referred to as *cloudlet nodes*—that an initiator mobile device can connect to for computing services. In mobile cloudlet, users can exploit the benefits of cloud computing without long wireless local area network (WLAN) communication latency as computing resources reside on local devices. In this chapter, we examine the fundamental properties of mobile cloudlet that unfold *whether and when a mobile cloudlet can provide mobile application services*. Specifically, we investigate the *cloudlet node's lifetime and reachable time*. Traces and

231

mathematical analysis demonstrate that (1) intermittent connection between devices has little adverse effect on the optimal computing performance of mobile cloudlet in the long run; (2) the ratio $E(T_C)/[E(T_I) + E(T_C)]$ indicates the connection likelihood of an initiator and a cloudlet node (i.e., reachability of the cloudlet node), where T_C and T_I are their contact and inter-contact time. We further derive upper and lower bounds on computing capacity and computing speed of a mobile cloudlet. An initiator can use both bounds to decide whether to off-load its task to remote clouds or local mobile cloudlets for better mobile application services.

6.1 Introduction

Mobile devices (such as smartphones and tablets) are becoming an inseparable part of our lives for convenient communication and entertainment. The number of smartphones in use worldwide reached 1.038 billion units during the third quarter of 2012, and smartphone users are expected to be over 2 billion by 2015 [2]. With the popularity of mobile devices, there is also an explosion of mobile applications in various categories, such as terrestrial navigation, email and web browsing, mobile games, mobile health care, mobile commerce, and social networking. The popularity of these mobile applications are evident through mobile app download centers, such as Apple's iTunes. This indicates that mobile devices are quickly becoming the dominant computing platform, which enables seamless work or entertainment for users regardless of user mobility.

Although mobile devices are rapidly gaining more computing power and memory resources [16], there is still a large gap in computing capabilities between mobile devices and desktops. Mobile systems are still limited in their resources (e.g., processor power, storage size, and battery life) and communications (e.g., bandwidth, connectivity, and security) [12]. Such resource scarceness significantly hinders the development of mobile applications and improvement of mobile service qualities. For instance, applications such as image processing for video games, speech synthesis, and augmented reality, demand high computational capacities thus their implementations for mobile devices are impeded.

Recently, this problem has been addressed through mobile cloud computing (MCC). MCC provides a service for resource constrained mobile devices to partition and off-load their computationally intensive and storage demanding jobs to the cloud with vast computational resources through high-speed and ubiquitous wireless connections. In MCC, computing-intensive mobile applications, such as video decoding, speech recognition, and augmented reality, can be off-loaded to the cloud for processing. Computation off-loading can save energy and improve the performance of mobile applications thereby overcoming the limited resource capacities of mobile devices. MCC can also enable mobile users to store and access the large data on the cloud through wireless networks, which can save data storage capacity and processing power on mobile devices.

MCC faces many challenges in the computing side such as application off-loading, security and privacy, and energy saving. Application off-loading is the main operation in MCC. Current research on off-loading can be categorized on the basis of entire application migration (e.g., [5,21]) and application partitioning (e.g., [7,15]). The entire application migration means that mobile devices off-load entire processing job to remote server nodes; application partitioning-based off-loading means that computationally intensive components of an application are separated statically or dynamically at compile time or runtime and partitions of the application are off-loaded to the cloud servers. Security issues in MCC are inherited from cloud computing [25], such as confidentiality and privacy, integrity, and availability. Security breaches may occur at different levels: network level, host level, and application level. Moreover, MCC is vulnerable for security attacks due to the open wireless communication medium and needs lightweight security algorithms due to the resource limitation of mobile devices. In addition, both mobile computation off-loading and data backup in MCC consumes the limited mobile device energy. In order to minimize the overall energy consumption, the decision between local processing and computation off-loading needs to be made carefully based on the amount of transferred data and computation, traffic pattern, and network bandwidth.

Besides challenges in the computation side, such as computation off-loading, MCC faces issues in wireless communication such as low bandwidth, service availability, and network heterogeneity [18,23]. The computational resources in the cloud are feasible only if the information exchange between the cloud and the mobile devices through wireless networks is fast, reliable, and secure. However, mobile devices utilize comparatively more intermittent and unreliable wireless communication with lower bandwidth.

Mobility among varied network environments intensifies communication deficiencies. Although there are various wireless technologies to support mobile users, such as 3G, 4G, LTE, WiMAX, and WiFi, it is still hard to guarantee always-on connectivity and high-quality mobile access in MCC.

Traditionally, MCC uses the *client-server communication model* [8,10] shown in Figure 6.1. The *remote cloud* provides data storage and computing service while mobile devices are clients to access the service through wireless networks, mainly WLAN and cellular network,

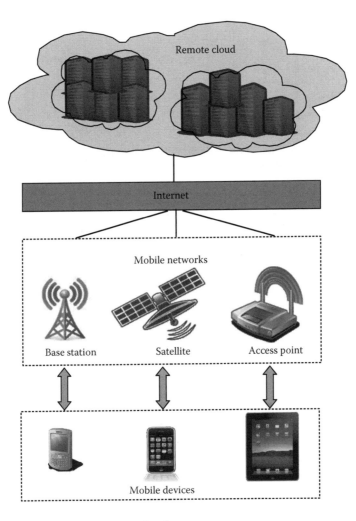

Figure 6.1   MCC uses remote cloud.

which could be unavailable or costly [16,21]. Specifically, although Wi-Fi has high data rate, Wi-Fi connections are intermittent. Cellular network provides the near-ubiquitous coverage, but cellular connection is known to suffer from very *long latency*, which may make mobile application off-loading expensive. For instance, a search query on a high-end smartphone with a 3G connection can take 3–10 s depending on location, device, and operator used. When 3G radio is not connected or only Enhanced Data rates for GSM Evolution connectivity is available, this delay can be doubled or even tripled. Moreover, off-loading mobile applications to remote cloud through 3G/4G connections may quickly *drain battery* [17]. In addition, cellular network is under significant pressure and likely to be overloaded due to the increasing mobile data traffic [1], which may incur long latencies and slow data transfers.

In order to overcome these drawbacks of accessing MCC through cellular and Wi-Fi networks, a *peer-to-peer communication model* for MCC [13,19,22] is proposed in light of the increasing memory and computational power of mobile devices [16]. As shown in Figure 6.2, *a group of nearby mobile devices can connect by WiFi or Bluetooth to form a mobile cloudlet*, in which mobile devices (referred to as cloudlet nodes) can be computing service providers as well as clients of the service. In many cases, processing mobile data (such as sensor logs and multimedia data) in-place and transferring it directly between smartphones would be more efficient and less susceptible to network limitations than offloading data and processing to remote servers. By dividing tasks among cloudlet nodes, an initiator mobile device could speed up computing and conserve energy. Users can get direct cloud computing access instantly through interactions among one another, eliminating the communication latency and data roaming charges introduced by cellular networks. Mobile cloudlet is especially appealing to users pursuing a common goal in group activities, such as multimedia sharing for audience at an event and language translation for a group of tourists in a foreign country.

Paper [21] points out that accessing remote cloud through wireless communication is costly because of long wide area network latencies. Rather than relying on a distant "cloud," the authors propose the use of *cloudlets* that instantiate software in real time on nearby computing resources using virtual machine technology. Huerta-Canepa and Lee [13] observed that mobile devices can be a virtual cloud computing provider because their pervasiveness means the increasing

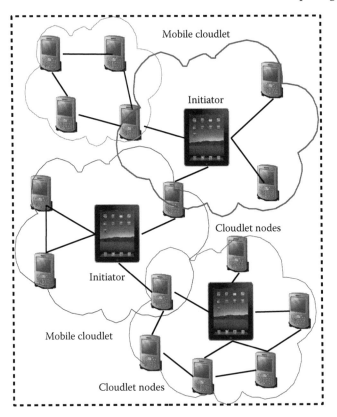

Figure 6.2   MCC uses mobile cloudlet.

availability of nearby devices; they are more powerful over time; they include different network interfaces allowing devices to communicate with each other (with no money cost); moreover they allow us to create communities in which we can execute shared tasks. Therefore, they proposed a virtual cloud computing platform, in which a context manager monitors the location and number of nearby devices. A P2P component utilizes an ad hoc discovery mechanism to track the status of the devices in the surroundings and then groups the nodes using a P2P scheme, allowing for better scalability and distribution of contents. Communication was performed using Ad Hoc WiFi between the mobile devices, and using an 802.11b/g compatible access point for accessing the servers. A Korean Optical Character Recognition that reads an image and scans for the Korean characters and then presents a Romanize version of them was developed for testing purposes.

Furthermore, paper [9] proposed a framework that uses local resources on mobile devices for computing when off-loading to remote clouds fails in low connectivity scenarios. Experiments are conducted in Bluetooth transmission and an initial prototype is also presented. The authors also discussed a preliminary analytical model to determine whether or not a speedup would be possible in off-loading. Marinelli [19] developed Hyrax, a platform derived from Hadoop that supports cloud computing on Android smartphones. Hyrax could facilitate sensor data applications (e.g., traffic reporting, sensor maps, and network availability monitoring) and multimedia applications (e.g., similar multimedia search, event summarization, and social networking). Hyrax allows client applications to conveniently utilize data and execute computing jobs on networks of smartphones and heterogeneous networks of phones and servers. A central server with access to each mobile device coordinates data and jobs and the smartphones communicate with each other on an isolated 802.11g network. Although the performance of Hyrax is poor for CPU-bound tasks, it is shown to tolerate node-departure and offer reasonable performance in data sharing. A distributed multimedia search and sharing application is implemented to qualitatively evaluate Hyrax.

In addition, paper [22] investigated the scenario that mobile device uses the available, potentially intermittently connected, computation resources of other mobile devices to improve its computational experience, for example, minimizing local power consumption and/or decreasing computation completion time. The authors proposed and implemented Serendipity on mobile devices to leverage the computing resources on a group of mobile devices through their frequent contacts in order to speed up computing and conserve energy. The authors also implemented a preliminary prototype of Serendipity on the Android platforms with two computationally complex applications (i.e., a face detection application and a speech-to-text application). Serendipity is shown to reduce job completion time comparing with executing locally.

The major concerns of using mobile cloudlet reside in the limited computing power of mobile devices and the unstable connections between mobile devices due to node mobility. Current research on mobile cloudlet shows the promise of overcoming these disadvantages of mobile cloudlet through prototype and framework implementations. Nevertheless, there still lacks theoretical analysis on the performance and feasibility of mobile cloudlet. Therefore, in this chapter, we investigate the fundamental questions in mobile cloudlet spectrum: what is the computing performance of mobile cloudlet under intermittent device connectivity, and whether and under what conditions mobile cloudlet can support mobile applications.

6.2   Models and Metrics

We consider a MCC network of n mobile nodes on a torus surface $\Omega_n = [0, \sqrt{n/\lambda}]$, where λ is the spatial density of mobile users. Suppose each mobile device has a transmission radius r. We assume that the mobility process of a node is stationary and ergodic that a node's location $X_i(\cdot)$ has uniform stationary distribution in the network area.

Without loss of generality, we assume that a mobile user needs to off-load a task to nearby mobile devices at time 0. Suppose the delay requirement of an initiator's task is τ, mobile devices that meet the initiator before the task expires have the potential to provide computing services, thus can form a mobile cloudlet for task computation. We assume that all nodes are willing to support cloudlet computing. Hence, *a mobile cloudlet is dynamically formed by the nodes that an initiator can connect to over a period of time* τ.

On one hand, mobile initiator can communicate with cloudlet nodes through *single-hop* wireless connection, which ensures short delay in task transfer and easy management of task distribution and retrieval. On the other hand, if an initiator employs mobile devices in *multi-hop* range, mobile cloudlet would potentially utilize more devices in a large area. However, multi-hop communication incurs longer delay and unreliable task dissemination and retrieval due to node mobility. Fesehaye et al. [11] show that when the maximum number of wireless hops in a cloudlet is larger than two, accessing cloudlet nodes incurs longer data transfer delay than directly accessing remote cloud through 3G/4G network. Hence, we only consider the more practical *single-hop mobile cloudlet*, which is formally defined as follows.

Definition 6.1 (*Single-Hop Mobile Cloudlet*) *For* $\tau \in \mathbb{R}_+$, *let* C_τ *be the mobile cloudlet for an initiator* v_i *with a task to compute within delay* τ. C_τ *is the set of nodes that* v_i *encounters within time* τ, *where cloudlet node* $v_j \in C_\tau$ *if and only if* $v_i \neq v_j$ *and there exists a link between* v_i *and* v_j *at a time* $0 \leq t \leq \tau$.

Apparently, node mobility affects mobile cloudlet's structure. Specially, how frequently nodes meet and how long they stay in contact affect the size and stability of a mobile cloudlet. In turn, contact and inter-contact time between nodes influence the computing capacity and performance of a mobile cloudlet as tasks can only be distributed and retrieved when there are communication links between an initiator and cloudlet nodes. Zhao et al. [24] find that the PDF of contact time T_C can be approximated by exponential distribution with parameter characterized by the ratio of average node speed to effective transmission range. Hence, in this chapter, we assume that T_C follows *exponential* distribution with parameter λ_C. Inter-contact time T_I has shown to exhibit exponential tail decay under many mobility models (such as random waypoint and Brownian motion) [3]. Analysis of a diverse set of mobility traces [14] also reveals that T_I follows a power law decay up to a characteristic time, beyond which T_I's distribution decays exponentially. For the simplicity of analysis, we assume that T_I has an *exponential* distribution with parameter λ_I. Note that our analysis and results can be easily extended to the case when T_C and T_I follow other distributions.

In order to evaluate the computing performance of mobile cloudlet, we explore metrics to represent the fundamental properties of mobile cloudlet that have direct implication on performance of mobile cloudlet. On the surface, node degree indicates the number of devices that an initiator can connect to at a given time and the potential computing resources in a mobile cloudlet. However, it cannot directly map to the performance of mobile cloudlet as mobile applications are executed by mobile cloudlet over a period of time. Hence, we propose the following two new metrics—cloudlet node's lifetime and reachable time.

Initially, an initiator has a task to compute in its single-hop mobile cloudlet within delay tolerance τ. Figure 6.3 shows the contact and

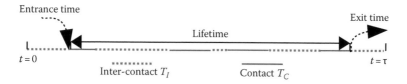

Figure 6.3   The contact and inter-contact process between an initiator and a cloudlet node.

inter-contact process between a cloudlet node and an initiator. The observation time is from time 0 when a task arrives at the initiator to time τ when the task needs to be finished. The connection process between two nodes alternates between contact and inter-contact. Figure 6.3 shows the case that the process is at inter-contact state at both $t = 0$ and $t = \tau$. The start point of a cloudlet node's first contact with the initiator is its entrance time to the initiator's mobile cloudlet and the end point of last contact is its exit time. Mobile cloudlet operations, that is, dispatching task to cloudlet nodes, computing task on cloudlet nodes, and retrieving task from cloudlet nodes, can only be performed after cloudlet nodes' entrance time and before their exit time. Thus, we define cloudlet node's lifetime as the duration from the entrance time to the exit time.

Definition 6.2 (Lifetime) *For any cloudlet node $v_j \in C_\tau$ for an initiator v_i, v_j's lifetime $LT(\tau) = t_{exit} - t_{entr}$, where its entrance time to C_τ*

$$t_{entr} \triangleq \inf_{0 \le t \le \tau} \{t : \|X_i(t) - X_j(t)\| \le r\},$$

and its exit time from C_τ

$$t_{exit} \triangleq \inf_{0 \le t \le \tau} \{t : \forall t' >= t \text{ and } t' \le \tau, \|X_i(t') - X_j(t')\| > r\}.$$

In an optimal situation, an initiator utilizes a cloudlet node's whole lifetime for computing. A cloudlet node receives tasks at its entrance time; then it can compute these tasks during its lifetime even when it is not in contact with the initiator; it finally sends back the tasks right before its exit time. Hence, the lifetime of a cloudlet node can be used to provide an *upper bound* on the computing capacity of mobile cloudlet.

Nevertheless, the task dissemination and retrieval can only be performed during the contact period of an initiator and a cloudlet node. The percentage of time that a cloudlet node is in contact with the initiator shows how likely the initiator can reach it. In order to study the reachability of cloudlet nodes and reliability of a mobile cloudlet, we define the *reachable time $RT(\tau)$ as the total contact duration between a cloudlet node and an initiator within time τ.*

By investigating cloudlet node's lifetime and reachable time, we can derive computing capacity of a mobile cloudlet, which is the average number of tasks that can be executed over time τ. Performance of mobile cloudlet can help mobile users to decide where to off-load mobile applications.

6.3   Properties and Performance Analysis of Single-Hop Mobile Cloudlet

In this section, we first mathematically analyze the properties of mobile cloudlet based on the modified alternating renewal process theory. Then, trace analysis and simulations are used to validate our analysis. Finally, results on mobile cloudlet properties will enable us to investigate the computing performance of mobile cloudlet, such as computing capacity and speed, and determine when mobile cloudlet is competent for executing mobile applications.

6.3.1   Lifetime of Cloudlet Node

We study a cloudlet node's lifetime based on the modified alternating renewal process (i.e., contact and inter-contact process between two devices) as follows:

Definition 6.3 *Let $\{(I_n, Z_n)\}_{n=0}^{\infty}$ be a bivariate stochastic process on a probability space (Ω, χ, P) such that $Z_0 = 0$ and $I_n = 1 - I_{n-1}$ for all $n \geq 1$, where $p_0 = P(I_0 = 0)$ and $p_1 = P(I_0 = 1)$ satisfying $p_0 + p_1 = 1$. Assume that, conditional on I_{n-1}, the $Z_n - Z_{n-1}$ for all $n \geq 1$ are mutually independent. Let the conditional distribution of Z_1 be $F_{\tilde{T}_C}^1$ $(F_{\tilde{T}_I}^1)$ if $I_0 = 1$ (0), and the distributions of $Z_n - Z_{n-1}$ conditioned on I_{n-1} be F_{T_C} (F_{T_I}) if $I_{n-1} = 1$ (0), for all $n \geq 2$. Distributions of $F_{\tilde{T}_C}^1$ and $F_{\tilde{T}_I}^1$ have density $\lambda_C[1 - F_{T_C}]$ and $\lambda_I[1 - F_{T_I}]$, respectively, where $\lambda_C^{-1} = \int_0^{\infty} F_{T_C}(dx)$ and $\lambda_I^{-1} = \int_0^{\infty} F_{T_I}(dx)$. Then, the point process characterized by $\{(I_n, Z_n)\}_{n=0}^{\infty}$, in which I_n is the point type and Z_n is the waiting time until the nth event, is an equilibrium modified alternating renewal process.*

Based on renewal theory, we deduce the following theorem on the expectation of a cloudlet node's lifetime.

Theorem 6.1 *The expected lifetime of a cloudlet node is shown in Equation 6.1, which can be approximated by $\tau - (1/\lambda_I)(1 - (\pi r^2/n/\lambda) + (\lambda_C/\lambda_I + \lambda_C))$ when τ is large.*

Proof: In equilibrium alternating renewal process, $\xi(t) = 1$ when two nodes are in contact at time t, $\xi(t) = 0$ otherwise.

(i) When $\xi(0) = \xi(\tau) = 1$, clearly $LT(\tau)$ equals to τ.

(ii) When $\xi(0) = 0$ and $\xi(\tau) = 1$, $LT(\tau)$ equals $\tau - \xi^1 \cdot 1_{\xi^1 < \tau}$, where $\xi^1 = \tilde{T}_I$ is the forward recurrence time of T_I.

(iii) When $\xi(0) = 1$ and $\xi(\tau) = 0$, $LT(\tau)$ equals $\tau - \xi^{N(\tau)} \cdot 1_{\xi^{N(\tau)} < \tau}$, where $N(\tau)$ is the number of contacts between two nodes, and $\xi^{N(\tau)} = \widehat{T_I}$ is the backward recurrence time of T_I.

(iv) When $\xi(0) = 0$ and $\xi(t) = 0$, $LT(\tau)$ equals $[\tau - (\xi^1 + \xi^{N(\tau)+1}) \cdot 1_{\xi^1 + \xi^{N(\tau)+1} < \tau}] \cdot 1_{\xi^1 < \tau}$, where $\xi^1 = \tilde{T_I}$ and $\xi^{N(\tau)+1} = \widehat{T_I}$.

Define $\pi_{ij}(\tau)$ as the equilibrium probability, given that $\xi(0) = i$, that $\xi(\tau) = j$. Let p_0 and p_1 denote $P(\xi(0) = 0)$ and $P(\xi(0) = 1)$, respectively. Because T_I and T_C are exponential random variables with parameters λ_I and λ_C, respectively, $\tilde{T_I}$ and $\widehat{T_I}$ have the same distribution as T_I and $\tilde{T_I} + \widehat{T_I}$ follows Erlang-2 distribution Erlang$(2, \lambda_I)$. Thus,

$$E(LT(\tau)) = \tau^2 \lambda_I e^{-\lambda_I \tau} \pi_{00}(\tau) p_0$$
$$+ \tau[1 + (\pi_{01}(\tau)p_0 + \pi_{10}(\tau)p_1 + \pi_{00}(\tau)p_0)e^{-\lambda_I \tau}]$$
$$- \frac{1}{\lambda_I}(1 - e^{-\lambda_I \tau})(\pi_{01}(\tau)p_0 + \pi_{10}(\tau)p_1 + 2\pi_{00}(\tau)p_0), \quad (6.1)$$

where
$$p_1 = \pi r^2 / (n/\lambda)$$
$$p_0 = 1 - p_1$$

The equilibrium probability $\pi_{ij}(\tau)$ can be derived based on Cox's Renewal Theory (Chapter 7.4) [6]: $\pi_{00}(\tau) = \beta + \gamma e^{-\beta \tau / \lambda_C}$, $\pi_{01}(\tau) = \gamma - \gamma e^{-\beta \tau / \lambda_C}$, $\pi_{10}(\tau) = \beta - \beta e^{-\beta \tau / \lambda_C}$, and $\pi_{11}(\tau) = \gamma + \beta e^{-\beta \tau / \lambda_C}$, where $\beta = \lambda_C / (\lambda_I + \lambda_C)$ and $\gamma = \lambda_I / (\lambda_I + \lambda_C)$. When τ is large, $e^{-\lambda_I \tau}$ and $e^{-(\lambda_I + \lambda_C)\tau}$ approach 0,

$$E(LT(\tau)) \approx \tau - \frac{1}{\lambda_I}\left(1 - \frac{\pi r^2}{n/\lambda} + \frac{\lambda_C}{\lambda_I + \lambda_C}\right) \quad (6.2)$$

To better understand $E(LT(\tau))$, we use numerical analysis of $E(LT(\tau))$ to show how $E(LT(\tau))$ changes with τ in Figure 6.4. We set $p_0 = 0.9$, $p_1 = 0.1$, $\lambda_I = 0.0001$, and $\lambda_C = 0.01$. Parameters λ_I and λ_C are set so that $E(T_I)$ and $E(T_C)$ approximately equal to 2.8 h and 1.7 min, respectively, that is, intermittent device connectivity with very short contacts. Figure 6.4 shows that when $\tau > 4 \times 10^4$, $E(LT(\tau))$ grows linearly with slope 1, which is consistent with Equation 6.2. When τ is small ($0 < \tau < 1000$), the close-up figure shows that $E(LT(\tau))$ is mainly influenced by τ^2.

Remark 6.1 *When τ is small, $E(LT(\tau))$ exhibits quadratic growth; as τ increases, $E(LT(\tau))$ grows linearly with slope 1 and the gap between τ*

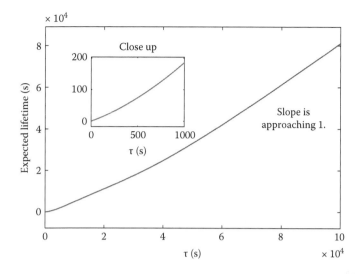

Figure 6.4   Expected lifetime of a cloudlet node grows linearly with slope 1 when τ is large.

and $E(LT(\tau))$ is a constant $(1/\lambda_I)\big(1 - (\pi r^2/(n/\lambda)) + (\lambda_C/(\lambda_I + \lambda_C))\big)$. This indicates that for application with long delay tolerance τ, intermittent connectivity between an initiator and cloudlet nodes has a small constant negative effect on the optimal computing performance, which is achieved when cloudlet nodes compute tasks for an initiator throughout their lifetimes. ∎

6.3.2   Reachable Time of Cloudlet Node

The reachable time of a cloudlet node equals to the total connection time between two nodes within time τ, which depends on not only distributions of contact and inter-contact time but also the number of contacts within time τ.

Theorem 6.2 *In a homogeneous network with uniform node distribution, if T_C and T_I follow exponential distributions with parameters λ_C and λ_I, respectively, the expected reachable time of a cloudlet node*

$$E(RT(\tau)) = \frac{\lambda_I \tau}{\lambda_I + \lambda_C} + \frac{\lambda_C p_1 - \lambda_I p_0}{(\lambda_I + \lambda_C)^2}(1 - e^{-(\lambda_I + \lambda_C)\tau}), \tag{6.3}$$

where
$$p_1 = \pi r^2/(n/\lambda)$$
$$p_0 = 1 - p_1.$$

Proof: M.H. Rossiter [20] derived the sojourn time distribution in on state for a two-state alternating renewal system by applying Laplace transform and double Laplace transform. For an equilibrium alternating renewal process, the Laplace transform of the expected sojourn time in on state conditioning on $I_0 = 0$

$$L(E[\alpha_0(\tau)]; s) = \frac{[1 - F_{T_C}(s)]F_{\tilde{T}_I}(s)}{s^2[1 - F_{T_C}(s)F_{T_I}(s)]}, \tag{6.4}$$

while conditioning on $I_0 = 1$

$$L(E[\alpha_1(\tau)]; s) = \frac{1 - F_{\tilde{T}_C}(s) - (F_{T_C} - F_{\tilde{T}_C})(s)F_{T_I}(s)}{s^2[1 - F_{T_C}(s)F_{T_I}(s)]}, \tag{6.5}$$

where

$L\{\cdot; s\}$ represents the Laplace transform

$F_X(s)$ is the Laplace transform of random variable X, $X = T_C$, T_I, that is, $F_X(s) = \int_0^\infty e^{-sx} F(dx)$

For exponential random variable T_C (T_I), its forward recurrence time $F_{\tilde{T}_C}$ ($F_{\tilde{T}_I}$) also has exponential distribution with parameter λ_C (λ_I) because of the memoryless property of exponential random variable. Then, $F_{T_C}(s) = F_{\tilde{T}_C}(s) = \lambda_C/(s + \lambda_C)$ and $F_{T_I}(s) = F_{\tilde{T}_I}(s) = \lambda_I/(s + \lambda_I)$. Thus,

$$L(E[\alpha_1(\tau)]; s) = \frac{1 - (\lambda_C/(s + \lambda_C))}{s^2[1 - (\lambda_C/(s + \lambda_C))(\lambda_I/(s + \lambda_I))]},$$

and $L(E[\alpha_0(\tau)]; s) = \lambda_I/(s + \lambda_I)L(E[\alpha_1(\tau)]; s)$. Performing inverse Laplace transform, we have

$$E[\alpha_0(\tau)] = \frac{\lambda_I \tau}{\lambda_I + \lambda_C} + \frac{\lambda_I}{(\lambda_I + \lambda_C)^2}(e^{-(\lambda_I + \lambda_C)\tau} - 1), \tag{6.6}$$

$$E[\alpha_1(\tau)] = \frac{\lambda_I \tau}{\lambda_I + \lambda_C} + \frac{\lambda_C}{(\lambda_I + \lambda_C)^2}(1 - e^{-(\lambda_I + \lambda_C)\tau}). \tag{6.7}$$

In our homogeneous network model, $p_0 = \pi r^2/(n/\lambda)$ and $p_1 = 1 - p_0$. Substituting them into $E(RT(\tau)) = E(\alpha_0(\tau))p_0 + E(\alpha_1(\tau))p_1$ completes our proof.

Remark 6.2 *The increase rate of $E(RT(\tau))$ shows that the mean reachable time within time τ is mainly determined by the ratio $E(T_C)/(E(T_I)+E(T_C))$ when τ is large. Nodes that meet an initiator frequently and have long contact time have high connection likelihood and can provide reliable computing services while still support mobility of the initiator. The initiator can estimate the ratio $E(T_C)/(E(T_I) + E(T_C))$ based on its contact histories and use it as an indicator for whether an encountered node is suitable for providing reliable mobile cloudlet service.* ∎

6.3.3 Trace Evaluations of Mobile Cloudlet Properties

As shown by our analysis in the previous section, cloudlet nodes' lifetime and reachable time are determined by contacts and inter-contacts between two nodes, which have been studied using mobility traces in mobile wireless networks. Hence, we validate our analysis on mobile cloudlet properties using mobility traces.

6.3.3.1 Mobility Traces

Mobility traces record mobile users' access to base stations or access points (i.e., infrastructure-based traces), or GPS locations (i.e., GPS-based traces), or contact and inter-contact time (i.e., direct contact-based traces). Because mobile cloudlet exploits contacts among nodes for computing, we choose the direct contact-based traces. Moreover, as mobile cloudlet is promising for a social group sharing common tasks, mobility traces of users in social groups are preferred. Therefore, we select the *Cambridge/haggle2009* dataset [4] that includes several traces of Bluetooth sightings by groups of users carrying small devices (iMotes) for several days in office and conference.

In Cambridge/haggle2009 data collection, experiment 2 distributed iMotes to 19 graduate students from the System Research Group at University of Cambridge for around 5 days in 2005. Number of contacts, contact, and inter-contact time among nodes were collected. Only 12 iMotes were used to produce trace file *Exp2*, while others were discarded because of hardware resets. Similarly, experiment 3 distributed iMotes to 50 students attending the student workshop at the IEEE Infocom Conference in Grand Hyatt Miami from March 7 to 10, 2005. Only 41 iMotes delivered useful contact information for trace file *Exp3*.

Exp2 and Exp3 represent node contact on campus and in conference environments, respectively. In both scenarios, mobile users are likely to work on a common task due to their common social activities (i.e., working in the same lab and attending the same conference). Nearby mobile users can create computing communities in which mobile devices can collaboratively execute shared tasks. Thus, properties of mobile cloudlet extracted from these two trace files can characterize the real mobile cloudlet system.

6.3.3.2   Lifetime of Cloudlet Node in Traces

Based on Definition 6.2, node v_j's lifetime in a mobile cloudlet for initiator v_i is from v_i and v_j's first contact to their last contact within time τ. Figure 6.5 shows the average $LT(\tau)$ in traces Exp2 and Exp3. $LT(\tau)$ increases as τ increases. Lifetime increases slowly when $80,000 < \tau < 132,000$ (about 14 h period) in trace Exp2 and when $90,000 < \tau < 129,000$ (about 11 h period) in trace Exp3. This is probably because users have little contact during nights, which is also observed in cloudlet node's reachable time in Figure 6.6.

Remark 6.3 *Figure 6.5 agrees with our theoretical results that a node's maximum lifespan for service increases linearly with rate 1 when τ is large. This implies that for delay tolerant application (i.e., large τ), the optimal computing performance of mobile cloudlet—achieved by exploiting mobile nodes' maximum lifespan for computing service—is hardly influenced by intermittent connections between mobile nodes.* ∎

6.3.3.3   Reachable Time of Cloudlet Node in Traces

The reachable time $RT(\tau)$ of a cloudlet node is its total contact duration with the initiator within time τ. Figure 6.6 shows the average reachable time in traces Exp2 and Exp3, which are piecewise linear functions. The piecewise linearity is due to different mobility patterns of users at different time (daytime and night time). For instance, in trace Exp2, students are working in the lab during $50,000 < \tau < 80,000$ (about 8 h period), producing long contact time and short inter-contact time. This leads to high growth rate of reachable time, as shown in Figure 6.6a. On the other hand, they have short contacts and long inter-contacts during off time, which lead to small growth rate of reachable time. Examining Figure 6.6 more closely, we discover that the increase rate depends on the average contact time and inter-contact time between two nodes over the corresponding period of time. For instance, the slope of segment $\tau \in [44,400, 85,200]$ in

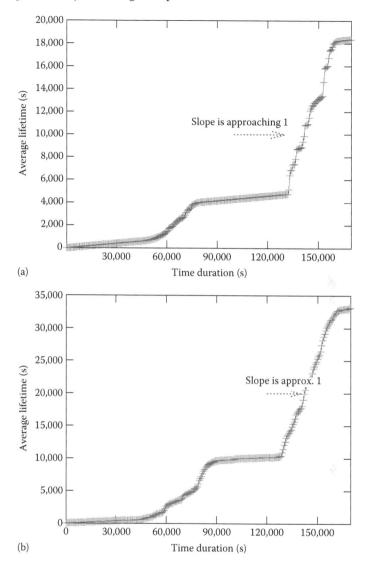

Figure 6.5   Average lifetimes of cloudlet nodes increase approximately linearly with τ when time τ is large: (a) Exp2 and (b) Exp3.

Figure 6.6b is 0.01882, which is very close to the average T_C/(average T_C + average T_I) (equal to 0.01877) during $44,400 < \tau < 85,200$.

Remark 6.4 *Analysis results of traces Exp2 and Exp3 validate our theoretical analysis. Cloudlet nodes' lifetime and reachable time increase linearly with τ when τ is large and the increase rates are approximate 1 and*

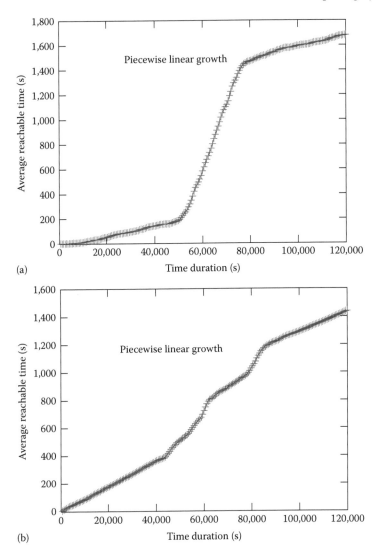

Figure 6.6   Average reachable times of cloudlet nodes are piecewise linear functions of time τ with slope depending on contact and inter-contact time: (a) Exp2 and (b) Exp3.

$E(T_C)/[E(T_C) + E(T_I)]$, *respectively. The growth rate of reachable time is varying according to users' mobility patterns, which indicates the time-varying connection likelihood between an initiator and a mobile device. If an initiator can connect to devices with high likelihood, it could receive omnipresent and reliable mobile cloudlet service.*   ∎

6.3.4 Computing Performance of Single-Hop Mobile Cloudlet

The amount of computation that the cloudlet nodes can provide for the initiator not only depends on the computing capabilities of cloudlet nodes and how the task is partitioned for parallel processing but also depends on the cloudlet node's lifetime and reachable time. Evaluating the computing capability of mobile processor and designing application partition schemes [7] are beyond the scope of this chapter. In this chapter, we simply assume that computing speed of each device is a constant V, and the initiator can partition the task into M subtasks that can be computed on cloudlet nodes in parallel. The data sizes of each subtask before and after computing are D_d^i and D_r^i ($1 \leq i \leq M$), respectively.

Assume a total of B Hz spectrum is shared by all nodes and each node has a fixed transmission power P. The noise N—including ambient and interference noise—is constant everywhere in the network. We characterize the wireless link using a pass loss model with attenuation exponent $\alpha \geq 2$. The capacity of a wireless link is $B\log_2(1 + (P/N)d^{-\alpha})$, where d is the Euclidean distance between the sender and the receiver. Assume that advanced error control coding is used such that the available link bandwidth is equal to its capacity. For a task contains D_d^i ($1 \leq i \leq M$) bits, the transmission time of dispatching a task is

$$0 < t_d = \frac{D_d^i}{B\log_2(1 + (P/N)d^{-\alpha})} \leq \frac{D_d}{B\log_2(1 + (P/N)r^{-\alpha})}, \qquad (6.8)$$

where $D_d \triangleq \sum_{1 \leq i \leq M} D_d^i$. The transmission time of retrieving a task is

$$0 < t_r = \frac{D_r^i}{B\log_2(1 + (P/N)d^{-\alpha})} \leq \frac{D_r}{B\log_2(1 + (P/N)r^{-\alpha})}, \qquad (6.9)$$

where $D_r \triangleq \sum_{1 \leq i \leq M} D_r^i$. Then the total transmission time is

$$0 < t_d + t_r \leq \frac{D_d + D_r}{B\log_2(1 + (P/N)r^{-\alpha})} \triangleq \rho. \qquad (6.10)$$

If tasks are computed on cloudlet nodes during their whole lifetime, the optimal mobile cloudlet capacity is achieved. If an initiator only employs cloudlet nodes when they are in contact, that is, the computing times equal to reachable times minus transmission times, we have a lower bound on the computing capacity of a mobile cloudlet.

Based on this methodology and results in previous subsections, we have the following theorem on the computing capacity of a mobile cloudlet.

Theorem 6.3 *In a single-hop mobile cloudlet C_τ, its expected computing capacity is upper bounded by $C_{C_\tau}^u$ and lower bounded by $C_{C_\tau}^l$ in Equations 6.11 and 6.14, respectively.*

Proof: Clearly, the computing capacity of mobile cloudlet C_τ satisfies

$$E(C_{C_\tau}) < C_{C_\tau}^u \triangleq E(\mathcal{N}(\tau))E(LT(\tau))S, \tag{6.11}$$

where

$E(LT(\tau))$ can be found in Equation 6.1

$E(\mathcal{N}(\tau))$ is the cloudlet size, equal to $(n-1)\left[1 - (1 - (\pi r^2/n/\lambda))e^{-\lambda_I \tau}\right]$

If an initiator only employs nodes in contact, a node's total computing time over time τ is

$$CT(\tau) = RT(\tau) - (p_0 + p_1\pi_{10})\sum_{i=1}^{N(\tau)}\overline{CT^i} - p_1\pi_{11}P(T_C < \tau)\sum_{i=1}^{N(\tau)+1}\overline{CT^i}, \tag{6.12}$$

where

$\overline{CT^i} \triangleq T_C^i \cdot 1_{T_C^i < \rho_{sum}} + \rho_{sum} \cdot 1_{T_C^i \geq \rho_{sum}}$, $\pi_{i,j}$ $(i,j = 0,1)$ are the equilibrium probabilities

$N(\tau)$ is the number of renewals within time τ

Denote $N^0(\tau) = N(\tau)|\{I_0 = 0\}$ and $N^1(\tau) = N(\tau)|\{I_0 = 1\}$. According to the renewal equation for modified renewal process,

$$E(N^0(\tau)) = F_{\tilde{T}_I}(\tau) + \int_0^\tau E(N^0(\tau))dF_{T_I+T_C}(s),$$

$$E(N^1(\tau)) = F_{\tilde{T}_C}(\tau) + \int_0^\tau E(N^1(\tau))dF_{T_I+T_C}(s).$$

Using Laplace transform, we have

$$E(N^0(\tau)) = \gamma\lambda_C\tau + \gamma^2(1 - e^{-(\lambda_I+\lambda_C)\tau}),$$

$$E(N^1(\tau)) = \beta\lambda_I\tau + \beta^2(1 - e^{-(\lambda_I+\lambda_C)\tau}),$$

where
$$\beta = \lambda_C/(\lambda_I + \lambda_C)$$
$$\gamma = \lambda_I/(\lambda_I + \lambda_C)$$

Subsequently,

$$E(N(\tau)) = E(N(\tau)|I_0 = 0)p_0 + E(N(\tau)|I_0 = 1)p_1,$$

$$= \frac{\lambda_I \lambda_C \tau}{\lambda_I + \lambda_C} + \frac{p_0 \lambda_I^2 + p_1 \lambda_C^2}{(\lambda_I + \lambda_C)^2}(1 - e^{-(\lambda_I + \lambda_C)\tau}), \qquad (6.13)$$

where
$$p_1 = \pi r^2/(n/\lambda)$$
$$p_0 = 1 - p_1$$

The computing capacity of C_τ satisfies

$$C_{C_\tau} \geq \sum_{j=1}^{N(\tau)} CT^j(\tau)S.$$

Therefore, $E(C_{C_\tau})$ is lower bounded by

$$C_{C_\tau}^l \triangleq E(N(\tau))S\left\{E(TCT(\tau)) - E(\overline{CT})*\right.$$
$$\left.\left[p_1 \pi_{11} F_{T_C}(\tau)(E(N(\tau)) + 1) + (p_0 + p_1\pi_{10})E(N(\tau))\right]\right\}, \qquad (6.14)$$

where $E(\overline{CT}) = E(T_C \cdot 1_{T_C < \rho}) + \rho P(T_C \geq \rho) = \frac{1 - e^{-\lambda_C \rho}}{\lambda_C}$.

Figure 6.7 shows the numerical results of Theorem 6.3 by setting $\rho = 0.1$ s (typical packet transmission time in mobile wireless networks), $n = 10$, T_I and T_C are exponential random variables with $\mu_{T_I} = 83$ min, and $\mu_{T_C} = 17$ min. This parameter setting reflects a scenario with 10 students in a team or 10 colleagues at a conference working on the same project. They meet for about 17 min between classes or conference sessions to compute a common task by sharing computing resources on their mobile devices. Two curves in Figure 6.7 divide MCC into three categories: MCC relying on (1) remote cloud, (2) remote cloud or mobile cloudlet, (3) mobile cloudlet. When this group of users need to execute a task with computational demand C and delay requirement τ, if $C \geq C_{C_\tau}^u$, they should off-load the task to remote cloud; if $C_{C_\tau}^l < C < C_{C_\tau}^u$, they can use remote cloud or mobile

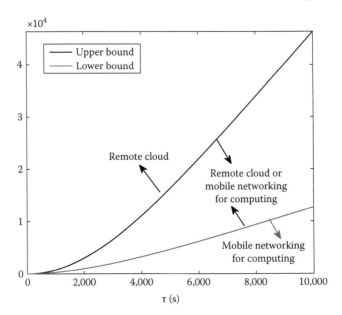

Figure 6.7   Bounds on computing capacity of single-hop mobile cloudlet.

cloudlet; if $C \le C_{C_\tau}^l$, they can implement mobile cloudlet to share task computation during their contacts.

The average computing speed of a mobile cloudlet is $E(C_{C_\tau})/\tau$. When $\tau \to \infty$, we have the long-term computing speed $CS = \lim_{\tau \to \infty} E(C_{C_\tau})/\tau$.

Theorem 6.4 *The long-term computing speed of a mobile cloudlet is upper bounded by $CS^u = (n-1)V$ and lower bounded by $CS^l = \frac{(n-1)V\lambda_I e^{-\lambda_C \rho}}{\lambda_C + \lambda_I}$.*

Proof: Based on Theorem 6.3, we have the upper bound of CS,

$$CS \le \lim_{\tau \to \infty} \frac{C_{C_\tau}^u}{\tau} = (n-1)V \triangleq CS^u. \tag{6.15}$$

Similarly, the lower bound of CS is $\lim_{\tau \to \infty} C_{C_\tau}^l/\tau$. As in modified renewal process,

$$\lim_{\tau \to \infty} \frac{E(N(\tau))}{\tau} = \frac{1}{E(T_C + T_I)} = \frac{\lambda_C \lambda_I}{\lambda_C + \lambda_I}.$$

Accordingly, we have

$$CS \geq \lim_{\tau \to \infty} \frac{C^l_{C_\tau}}{\tau} = \frac{\lambda_I e^{-\lambda_C \rho}}{\lambda_C + \lambda_I}(n-1)V \triangleq CS^l. \tag{6.16}$$

Remark 6.5 *The bounds on long-term computing speed can also be used by an initiator mobile device to decide where to off-load its task for computing service. Suppose the initiator has a task with computational demand C and delay requirement τ, if $C/\tau \geq CS^u$, the initiator needs to off-load its task to a remote cloud; if $C/\tau \leq CS^l$, the initiator can distribute its task to nearby devices for computing; otherwise, the initiator can choose either remote cloud or mobile cloudlet based on other constraints, such as battery life and quality of wireless communication.* ∎

6.4 Conclusion

In this chapter, we have explored the domain of mobile cloudlet in MCC through studying mobile cloudlet properties and computing performance. The intermittent connection has little negative effect on optimal performance of a mobile cloudlet in the long run. Furthermore, $E(T_C)/(E(T_C) + E(T_I))$ implies the connection likelihood of a cloudlet node to an initiator, thus can be used by the initiator to choose reliable cloudlet node. Based on cloudlet properties, we have also derived upper and lower bounds on the computing capacity and long-term computing speed of a mobile cloudlet. An initiator can use these bounds to decide whether to upload a task to remote clouds or utilize nearby mobile cloudlet. Future work on design and implementation of mobile applications on a mobile cloudlet system is expected to investigate the feasibility and performance of mobile cloudlet computing.

References

1. A. Aijaz, H. Aghvami, and M. Amani. A survey on mobile data offloading: Technical and business perspectives. *IEEE Wireless Communications*, 20(2):104–112, 2013.

2. S. Bicheno. Global smartphone installed base forecast by operating system for 88 countries: 2007 to 2017. Technical report, Strategy Analytics, Newton, MA, Oct. 2012.

3. H. Cai and D. Y. Eun. Toward stochastic anatomy of inter-meeting time distribution under general mobility models. In *Proceedings of the ACM MobiHoc*, Hong Kong, China, pp. 273–282, 2008.

4. A. Chaintreau, P. Hui, J. Scott et al. Impact of human mobility on opportunistic forwarding algorithms. *IEEE Trans. Mobile Comput.*, 6(6):606–620, 2007.

5. B.-G. Chun and P. Maniatis. Augmented smartphone applications through clone cloud execution. In *Proceedings of the 12th Conference on Hot Topics in Operating Systems, (HotOS'09)*. USENIX Association, Monte Verita, Switzerland, pp. 8, 2009.

6. D. R. Cox. *Renewal Theory*. Methuen & Co, London, United Kingdom, 1962.

7. E. Cuervo, A. Balasubramanian, D.-k. Cho et al. Maui: Making smartphones last longer with code offload. In *Proceedings of MobiSys*, San Francisco, CA, pp. 49–62, 2010.

8. H. T. Dinh, C. Lee, D. Niyato, and P. Wang. A survey of mobile cloud computing: Architecture, applications, and approaches. *Wireless Commun. Mobile Comput.*, 13(18):1587–1611, 2013.

9. N. Fernando, S. W. Loke, and W. Rahayu. Dynamic mobile cloud computing: Ad hoc and opportunistic job sharing. In *2011 Fourth IEEE International Conference on Utility and Cloud Computing (UCC)*, Melbourne, Australia, pp. 281–286, 2011.

10. N. Fernando, S. W. Loke, and W. Rahayu. Mobile cloud computing: A survey. *Future Gener. Comp. Syst.*, 29(1):84–106, 2013.

11. D. Fesehaye, Y. Gao, K. Nahrstedt, and G. Wang. Impact of cloudlets on interactive mobile cloud applications. In *The Proceedings of IEEE Enterprise Distributed Object Computing Conference (EDOC)*, Beijing, China, 123–132, 2012.

12. A. K. Gupta. Challenges of mobile computing. In *Second National Conference on Challenges & Opportunities in Information Technology (COIT-2008)*, Mandi Gobindgarh, India, pp. 86–90, 2008.

13. G. Huerta-Canepa and D. Lee. A virtual cloud computing provider for mobile devices. In *ACM Workshop on Mobile Cloud Computing & Services (MCS'10)*, San Francisco, CA, 6:1–6:5, 2010.

14. T. Karagiannis, J.-Y. Le Boudec, and M. Vojnovic. Power law and exponential decay of inter contact times between mobile devices. In *Proceedings of the ACM MobiCom*, Montreal, QC, Canada, pp. 183–194, 2007.

15. R. Kemp, N. Palmer, T. Kielmann, and H. Bal. Cuckoo: A computation offloading framework for smartphones. In M. Gris and G. Yang, eds., *Mobile Computing, Applications, and Services*, Vol. 76 of Lecture Notes of the Institute for Computer Sciences, Social Informatics and Telecommunications Engineering, pp. 59–79. Springer Berlin, Heidelberg, Germany, 2012.

16. E. Koukoumidis, D. Lymberopoulos, K. Strauss, J. Liu, and D. Burger. Pocket cloudlets. In *The 16th International Conference on Architectural Support for Programming Languages and Operating Systems (ASPLOS)*, Newport Beach, CA, pp. 171–184, 2011.

17. K. Kumar and Y.-H. Lu. Cloud computing for mobile users: Can offloading computation save energy? *Computer*, 43(4):51–56, 2010.

18. L. Lei, Z. Zhong, K. Zheng, J. Chen, and H. Meng. Challenges on wireless heterogeneous networks for mobile cloud computing. *IEEE Transactions on Wireless Communications*, 20(3):34–44, June 2013.

19. E. E. Marinelli. Hyrax: Cloud computing on mobile devices using MapReduce. Master's thesis, Carnegie Mellon University, Pittsburgh, PA, 2009.

20. M.H. Rossiter. The sojourn time distribution of an alternating renewal process. *Australian Journal of Statistics*, 31(1):143–152, 1989.

21. M. Satyanarayanan, P. Bahl, R. Caceres, and N. Davies. The case for VM-based cloudlets in mobile computing. *IEEE Pervasive Computing*, 8(4):14–23, 2009.

22. C. Shi, V. Lakafosis, M. H. Ammar, and E. W. Zegura. Serendipity: Enabling remote computing among intermittently connected mobile devices. In *Proceedings of ACM MobiHoc*, Hilton Head, CA, pp. 145–154, 2012.

23. Y. Xu and S. Mao. A survey of mobile cloud computing for rich media applications. *IEEE Trans. Wireless Commun.*, 20(3):46–53, June 2013.

24. M. Zhao, Y. Li, and W. Wang. Modeling and analytical study of link properties in multihop wireless networks. *IEEE Trans. Commun.*, 60(2):445–455, 2012.

25. M. Zhou, R. Zhang, W. Xie, W. Qian, and A. Zhou. Security and privacy in cloud computing: A survey. In *2010 Sixth International Conference on Semantics Knowledge and Grid (SKG)*, Ningbo, China, pp. 105–112, 2010.

CHAPTER 7

Software Piracy Control Framework in Mobile Cloud Computing Systems

Mazhar Ali, Muhammad Usman Shahid Khan, Assad Abbas,
and Samee U. Khan

CONTENTS

7.1   Introduction

The ubiquity of the cloud computing allows the mobile devices to connect and use the traditional cloud computing services [1]. However, unlike the normal computing machines, the mobile devices are resource constrained. The precincts of low processing power, less storage capacity, limited energy, and the capricious Internet connectivity do not allow the compute and storage mandating applications to run on mobile devices [2]. The aforementioned limitations served as the motivation for a new computing paradigm called mobile cloud computing (MCC) that enhances the abilities of mobile devices by offloading resource-intensive applications to the cloud. Mobile devices can now execute heavy compute and storage-intensive applications by using the computation and storage services of the cloud [3]. The MCC paradigm enables the users to access and manage their applications and data through mobile devices without the need to move to traditional computing machines.

257

The MCC possesses multifaceted advantages including (a) the ability to empower the mobile devices to perform compute intensive tasks, (b) boosting the mobile device performance by exploiting cloud computational capabilities, and (c) cutting down the energy consumption of mobile devices by delegating the computational tasks to cloud [4]. However, to benefit from the aforementioned capabilities of the MCC, specialized applications that support the MCC paradigm are required [5]. The reason for the development of the specialized applications is the inability of the traditional mobile applications to use the said characteristics of the MCC. To provide the aforementioned features, mobile application development models are used that off-load the mobile applications or the smartphone clone to the cloud [2]. Nevertheless, the off-loading of clone or application to the cloud may raise the software piracy issues [5]. The traditional mobile applications bind the application license mostly with the mobile device. In the MCC paradigm, when an application or the smartphone clone is off-loaded to the cloud, the device-dependent parameters are no longer visible to the cloud. The absence of both the device parameters and the environment makes the cloud unable to verify the application license [5].

The issue of software piracy in mobile applications needs an immediate attention of the industry and academia due to the fact that the use of mobile devices is escalating with the swift rates. The survey conducted by the International Data Corporation (IDC) [6] states that by year 2015 the number of mobile devices requiring access to the Internet will exceed the number of conventional computers. Likewise, the number of cloud supported devices is expected to be around one trillion by the end of year 2014 [7]. Moreover, the development of mobile applications is anticipated to outstrip the development of conventional applications within next 5 years [8]. As a result of exponential growth in the development and usage of mobile devices and applications, the MCC market is estimated to outdo the value of $10 billion in year 2015 [7]. However, the aforesaid statistics also bring the issue of application piracy up front with a severity level that never existed before.

According to the available statistics, Football Manager is the most pirated smartphone application. The Football Manager has 500 times more pirated copies than the legal copies [9]. Moreover, around 84% of the available iOS applications are affected by the piracy issues [9]. China that makes more than 25% of the total world's population has around 40% of pirated Android applications [10]. The piracy issues not only entail a financial loss to the application developers but may also increase the cost of cloud based services due to free usage. The pirated cloud application may utilize cloud services that can cost financial liabilities to the application providers. The year 2010 has observed 75% application downloads without payment that deprived application developers of 70% of the expected income [8]. Similarly, a research project has shown that the cloud enabled mobile applications can be run on cloud for free by manipulating the loopholes of the MCC applications [11]. The result of such happenings is a huge financial loss to not only the application developers but to the cloud service providers as well.

As mentioned earlier, the prevailing piracy control strategies do not solve the piracy issue in the MCC paradigm. Moreover, very little attention is paid to the piracy issue in the MCC by research community [5]. Therefore, it is the high time for the research community to focus on the piracy issues prevalent in MCC to cater the problem that not only involves financial concerns but also has ethical obligations. The aforesaid situation demands the research efforts to develop a piracy control framework for mobile cloud applications.

In this chapter, we propose a software piracy control framework for mobile cloud applications. The proposed framework considers the piracy control issue as one of the offshoots of access control. The software license ensures that the software is executed only by the party that is authorized or has been granted access to execute. In the same manner, the proposed framework gives access to the requesting user to execute the application on the cloud only if the user is authorized to do so. The license verification and access grant in the proposed framework is ticket-based where the credentials and parameters are verified by the possession of the valid ticket or otherwise. The ticket is issued for a specified period and the execution of the application after that time will require the acquisition of the new ticket.

7.2   Related Work

There are numerous strategies and products for license management in conventional computing paradigm. However, the research on license management in MCC has little been explored. In the following text we will explain few of the works that are based either on MCC or related to the cloud computing paradigm.

Khan et al. [5] presented a framework called pirax for piracy control in MCC environment. The pirax framework averts the unauthorized execution of mobile applications not only on a mobile device but also on cloud. The pirax is a node-locked licensing methodology that binds the device-dependent parameters to enforce the license. On the device side, the pirax uses International Mobile Equipment Identity (IMEI) to bond the application with the mobile device. The application provider generates a unique ID for the application, and concatenates IMEI of requesting user and the generated application ID. Subsequently, the concatenated result is passed through a hash function. The resulting hash code is termed as a license that is applicable only at the mobile device. The license is transmitted to the mobile device after encrypting with the public key of the mobile device. The license is validated before application execution by generating the hash of application ID and IMEI locally and comparing with the received license. The successful comparison results in execution of the application on mobile device. However, when application is executed on the cloud, the IMEI cannot be accessed because the smartphone clone is off-loaded to the cloud. Therefore, a request to the application provider is sent for providing cloud license for the application. The unique ID of the virtual machine (VM) that is assigned to the smartphone clone is also transmitted to the application provider. The application provider concatenates unique VM-ID, application ID, and previously generated license for mobile device. The result of the concatenation is passed through hash function to generate the cloud side license for the application. The same parameters are used at the cloud side to generate hash for validating the license received from the application provider. However, the pirax framework works only for those mobile application models that require the whole smartphone clone to be off-loaded to the cloud. Other models that require only parts of the application or mobile device are not handled by the pirax framework. Moreover, at the mobile device side IMEI is used that can be spoofed and the pirax does not deal with such hostile scenario.

The authors in [12] have presented GenLM that is token-based approach for license management in grid and cloud environment. GenLM does not bind the license to a node or user. The license is attached to the data files that need to go for computations on the remote site. The user calculates the hash of the input data files.

The hash of each of the input file is stored in a separate file that is termed as request token. The request token also contains the terms that are requested by the user from the application provider. Afterward, the token is signed by the user with X.509 certificate. The signed request token is sent to the GenLM server. The GenLM server entertains the service requests for all of the application providers for license generation and verification. The server verifies the signature on the request token and forwards the request to the policy plugin of the application provider from which the license is requested. The policy plugin evaluates the user request on the basis of application provider's business model and decides whether to grant license on the requested terms or not. The policy plugin also handles the billing matters of the license. If the license is granted, the GenLM server signs the request token with its X.509 certificate. After signature of the GenLM server, the request token becomes the license token that is sent to the user. The user submits the license token along with the input data files to the grid or the cloud. The job is only executed if the hash of the input files in the license token matches the hash of the input files generated at the compute site. Although the proposed approach handles the computations at the grid and cloud, it only caters the computational data. Moreover, the technique is not valid for mobile cloud applications because the technique computes hash over the data only.

The pirax framework that we highlighted in this section is the only proposed framework that tries to handle the piracy issue in the MCC paradigm. However, the pirax framework has its own limitations that we discussed in the preceding discussion. In the next section, we present our proposed software piracy control framework for the MCC environment. The proposed framework is ticket-based and is applicable to all mobile application models.

7.3   The Proposed Framework

The conventional node-locked phenomenon for enforcing piracy control does not prove to be fruitful in the MCC environment. The reason for the aforesaid fact is that the device-dependent parameters cannot be effectively accessed and employed in the cloud. The off-loading to the cloud is done only for application, part(s) of application, or the whole clone of the mobile device. All of the aforesaid models cannot access and utilize the device parameters in the same way as on the device itself. Therefore, in the proposed framework we use a ticket-based mechanism to validate the license and grant access for the execution of the mobile application in the cloud.

7.3.1 Entities

The following entities are involved in the proposed methodology.

User: The clients that use the mobile application are the users in the proposed methodology. The users will purchase the mobile application from the application provider. The users can execute the mobile application either at the mobile device or at cloud at different points of time. The license needs to be enforced at both of the sides.

Application provider: The application provider is the developer of the mobile application. The application provider's objective is to eliminate the pirated and illicit execution of the mobile applications. The mobile application needs to be executed only by the customer who purchased and has rights for execution. The business model of the application provider can only be successful and fruitful by returning what is due for the application provider. A License Granting and Verification Server (LGVS) at the application provider's end handles the users' request for granting license and execute permissions in the form of a ticket. The application either at device or cloud needs a ticket to verify the genuineness of the user license for subsequent execution.

The Cloud: The cloud provides computational services to the users. The users off-load their application to the cloud to take advantage of the computational power of the cloud. Moreover, the performance and energy consumption of the mobile device is enhanced by outsourcing the computations to the cloud. The execution of the off-loaded mobile applications to the cloud is managed by Access Granting Server (AGS) that is present at the cloud. The AGS at the cloud and the LGVS at the application provider's site work in coordination to control the illicit mobile application's execution.

7.3.2   The Proposed Methodology

In this section, we present the details of the proposed framework to prevent the illicit execution of mobile applications. The framework ensures that the application is executed only by the authorized user. Initially, the user contacts the application provider to purchase the application. The pricing and the license terms are agreed upon according to the procedure set by the application provider. Subsequently, the user installs the application on the mobile device. However, the execution of the mobile application needs the permission of the application provider. The same permission is needed to execute the application after off-loading to the cloud.

When the application needs to be executed on the mobile device, the first time execution is interrupted to acquire the ticket from the LGVS. The application extracts the IMEI of the mobile device. The IMEI is sent to LGVS along with the application ID, and the requesting user ID. The IMEI is encrypted with the public key of LGVS and subsequently signed by the private key of the mobile device. The encryption and signing are meant to prevent IMEI spoofing by unauthorized users and applications. The LGVS authenticates the identity of the user and application and checks whether the user has obtained a valid license or not. A locally maintained database contains the information about the issued licenses and user identities. After successful authentication, the LGVS prepares and sends the following message to the application.

$$Tic\_MD = \{\{\{IMEI \| Application\,ID \| validity\,period\}_{pub\_MD}\}_{pri\_LGVS}\}$$

where

Tic_MD represents the license for application execution on mobile device

$\|$ represents the concatenation operator

pub_MD is the public key of the mobile device

pri_LGVS is the private key of LGVS

The application at the mobile device verifies the LGVS signatures, extracts the application ID and IMEI and compares them with local values of the same parameters. The validity period is verified and upon success the mobile application is allowed to execute. At the mobile device, the strategy is more like a node-locked phenomenon.

However, at cloud side the process is different. At the device end, the
Tic_MD can be stored by the application and can be used during the
life cycle of the license. However, to execute application at the cloud,
the license is verified every time the application is executed.

Whenever a user wants to execute the mobile application at the
cloud or at the device, he/she contacts the LGVS of the corresponding
application provider. The user sends authentication credentials and
the application ID for which the user has obtained the license. The
LGVS after successful authentication checks whether the user has got
a valid license or not. The check is performed from a locally maintained database at LGVS. In case of a valid license, the LGVS sends a
message to the mobile user with the following contents:

$$M =_{\iota} \{\{nonce\}_{pub\_user}, \{AT\}_{pub\_AGS}\}$$

where
   AT is the access token
   pub_user is the public key of the user
   pub_AGS is the public key of the AGS
   $\{nonce\}_{pub\_user}$ is the nonce encrypted with the public key of the user
   $\{AT\}_{pub\_AGS}$ is the AT encrypted with the public key of AGS

The AT is further comprised of the following attributes:

• Requesting user ID

• Token validity period

• Application ID

• Nonce

The mobile user decrypts the nonce. The AT cannot be decrypted
by the mobile user as it is encrypted by the public key of AGS.
To execute the application at the cloud, the mobile user has to
access the cloud through the AGS. The mobile user prepares the following message to send to the AGS along with the authentication
credentials.

$$M_{AGS} = \{\{\{nonce\}_{pub\_AGS}\}_{pri\_user}, \{AT\}_{pub\_AGS}\}$$

In the earlier given message, the *nonce* is signed by the sending user
after encrypting with the public key of the AGS. The AGS verifies the

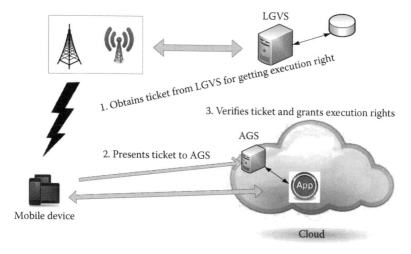

Figure 7.1   General architecture for proposed framework.

user signature and decrypts the nonce. Subsequently, it compares the nonce received by the mobile user and the nonce received from LGVS. The aforesaid step authenticates the requesting user. Moreover, the AGS verifies the validity of the license by extracting parameters from the *AT*. The AGS verifies the application ID received in the *AT* by comparing with the application ID of the off-loaded application. The validity period of the token is also verified. If the license is valid, the AGS grants access to the user for executing the mobile application at the cloud. The request is denied otherwise. Figure 7.1 shows the general architecture of the proposed framework while Figure 7.2 depicts the workflow of the proposed framework for executing the application at the cloud.

7.4   Discussion

The proposed framework has two courses of actions to deal with the application execution at the mobile device and at the cloud. The process adopted at the device end is much like a node-locked phenomenon. However, at the cloud side the ticket needs to be issued by the LGVS and verified by the AGS each time the application is off-loaded to the cloud. It is noteworthy that the proposed framework is flexible enough to work with all of the models for application off-loading to the cloud. The framework works with the complete

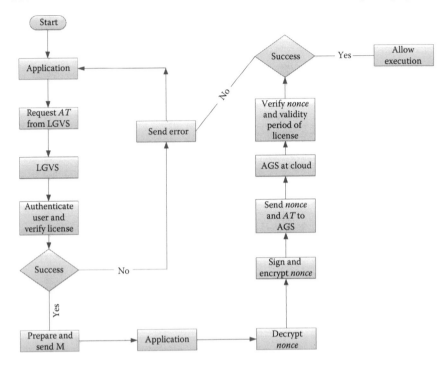

Figure 7.2   Workflow of proposed framework for mobile application execution at cloud.

clone or only application or part(s) of application off-loaded to the cloud. The proposed framework operates on the basis of ticket issuance. The ticket is issued by the application provider and verified by the AGS at the cloud. The validity period and the users are authenticated each time the request for application execution is initiated.

The licenses are generated by the LGVS that is proposed to be managed by the application provider. The communication of vital parameters is secured by use of public/private keys that can be managed by X509 certificates. Therefore, the proposed framework also considers the security aspect to provide better services to both the application providers and the users.

7.5   Conclusions

The mobile applications are off-loaded to cloud to enhance the mobile device's performance and take advantage of the cloud's computational capabilities. However, off-loading the mobile application to the cloud raises application piracy concerns. We proposed a software piracy control framework in MCC systems. The framework approaches the issue from the access control perspective and is based on tickets for verification of a valid license. An LGVS verifies the user and grants a ticket for certain time. The AGS after verification of the ticket issued by LGVS, grants execution rights to mobile application at the cloud. The cryptographic parameters are used in the proposed framework to obtain the verification services and keep the messaging secure.

References

1. A. N. Khan, M. L. M. Kiah, S. U. Khan, and S. A. Madani, Towards secure mobile cloud computing: A survey, *Future Generation Computer Systems*, 29(5), 2013, 1278–1299.

2. A. R. Khan, M. Othman, S. A. Madani, and S. U. Khan, A survey of mobile cloud computing application models, *IEEE Communications Surveys & Tutorials*, 16(1), 2013, 393–413.

3. A. N. Khan, M. L. M. Kiah, M. Ali, S. A. Madani, and S. Shamshirband, BSS: Block-based sharing scheme for secure data storage services in mobile cloud environment, *The Journal of Supercomputing*, 70(2), 2014, 946–976.

4. A. N. Khan, M. L. M. Kiah, S. U. Khan, and S. A. Madani, A study of incremental cryptography for security schemes in mobile cloud computing environments, In *2013 IEEE Symposium on Wireless Technology and Applications (ISWTA)*, Kuching, Malaysia, 2013, pp. 62–67.

5. A. R. Khan, M. Othman, M. Ali, A. N. Khan, and S. A. Madani, Pirax: Framework for application piracy control in mobile cloud environment, *The Journal of Supercomputing*, 68(2), 2014, 753–776.

6. International Data Corporation, Worldwide mobile phone forecasts and analyses, http://www.idc.com/getdoc.jsp?containerId=prUS23028711, Nov. 2014, accessed January 2015.

7. P. A. Cox, Mobile cloud computing—Devices, trends, issues, and the enabling technologies, available at http://public.dhe.ibm. com/software/dw/cloud/library/cl-mobilecloudcomputing- pdf.pdf, Nov. 2014, accessed February 2015.

8. Worldwide application development and deployment, http:// www.idc.com/getdoc.jsp?containerId=IDC_P1600, Nov. 2014, accessed January 2015.

9. Dear Android: A 9:1 piracy rate for games is not good enough, http://www.wired.co.uk/news/archive/2012–05/02/android market-game-piracy, Nov. 2014.

10. S. Hanna, L. Huang, E. Wu, S. Li, C. Chen, and D. Song, Juxtapp: A scalable system for detecting code reuse among android applications, In *Detection of Intrusions and Malware, and Vulnerability Assessment*, U. Flegel, E. Markatos, and W. Robertson (eds.), Springer Berlin, Heidelberg, Germany, 2013, pp. 62–81.

11. V. Tendulkar, R. Snyder, J. Pletcher, K. Butler, A. Shashidharan, and W. Enck, Abusing cloud-based browsers for fun and profit, In *Proceedings of the 28th Annual Computer Security Applications Conference*, Orlando, FL, 2012, pp. 219–228.

12. M. Dalheimer and F.-J. Pfreundt, Genlm: License management for grid and cloud computing environments, In *Ninth IEEE/ACM International Symposium on Cluster Computing and the Grid*, Shanghai, China, 2009, pp. 132–139.

CHAPTER 8

Cloud Radio Access Networks in Mobile Cloud Computing Systems

Yegui Cai, F. Richard Yu, and Shengrong Bu

CONTENTS

8.1   Introduction

Recently, as a new information technology (IT) paradigm, cloud computing has become one of the hottest topics in both academia and industry. Cloud computing is a model for enabling on-demand access to a shared pool of configurable resources (e.g., servers, storage, applications, services, etc.). The essential characteristics of cloud computing include on-demand self-service, broad network access, resource pooling, rapid elasticity, and measured service [26]. Several service models are supported, including cloud software as a service, cloud platform as a service, and cloud infrastructure as service [24]. Cloud computing has attracted significant attention, and several commercial clouds, including Amazon EC2, Microsoft Azure, and Google App Engine, have been providing services to users.

Cloud computing will have profound impacts on the design and operation of wireless networks. On one hand, with recent advances of wireless mobile communication technologies and devices, more and more end users access cloud computing systems via mobile devices, such as smart phones and tablets. The integration of cloud computing into the mobile environment enables *mobile cloud computing* (MCC), which is widely considered as a promising mobile computing paradigm with huge market [11,38]. MCC enables off-loading the computing power and data storage requirements from mobile devices into the powerful computing platforms in the cloud, bridging the gap between the increasing computing demands and the traditional mobile technologies with limited computing, storage, and energy resource in mobile devices [11].

On the other hand, the powerful computing platforms in the cloud can be beneficial to radio access networks (RAN) as well (in addition to mobile end users), which leads to a novel concept of *cloud radio access networks* (C-RAN) [4,9,30]. Unlike the existing cellular networks, where computing resources for baseband processing are located at each cell site, in C-RAN, the computing resources are located in a central wireless network cloud with powerful computing platforms. This transition from distributed to centralized infrastructure for baseband processing can have significant benefits: saving the operating expenses due to centralized maintenance; improving network performance due to advanced coordinated signal processing techniques; reducing energy expenditure by exploiting the load variations [4,9,30].

Although some excellent works have been done to study cloud computing for both end users and access networks, these two important areas have traditionally been addressed separately in the literature. However, as shown in the following, it is necessary to consider these two advanced technologies together to provide better services in next-generation wireless networks. Therefore, we jointly study C-RAN in MCC systems so as to improve end-to-end network performance. The motivations behind our work are based on the following observations.

- From end-to-end applications' perspective, both C-RAN and over-the-top (OTT) service provider cloud are parts of the whole system. The experience in end-to-end applications (e.g., transmission control protocol [TCP]-based applications) indicates that the optimized performance in one segment of the whole system does not guarantee the end-to-end improvement if the other segment is ignored in the optimization [3].

- It is well known that, while TCP performs relatively well over wired links, its performance degrades over wireless links due to the scarce bandwidth, high bit error rate, and user mobility [34]. In addition, networking has become a bottleneck that has a significant impact on the quality of cloud services [36]. Therefore, the characteristics of C-RAN should be carefully considered in MCC systems.

- Recent studies in cross-layer designs show that optimizing lower layer's performance (e.g., physical layer throughput) does not necessarily benefit quality of service (QoS) at upper layers [21]. From a user's point of view, QoS at upper layers (e.g., TCP throughput) is more important than that at other layers.

To the best of our knowledge, the study of C-RAN in MCC systems for next-generation wireless networks has not been addressed in previous works. The distinct features of this work are as follows.

- We consider how to dynamically configure C-RAN to enhance MCC services' performance in a holistic framework. For C-RAN, we study the topology configuration and rate allocation problem; for MCC, as a case study, we consider mobile search engine services on top of TCP connections, which has critical requirements on per user response latency. In this work, we optimize the end-to-end TCP throughput performance of MCC users in next-generation cellular networks.

- Despite the potential benefits brought by C-RAN, one of the major challenges in C-RAN is that the channel state information (CSI) is inaccurate due to the delay in obtaining and transmitting such information [30]. For instance, in Long-Term Evolution-Advanced, the standard interface for inter-base station communications, $X2$, is designed to allow a latency of 20 ms for control plan messages, and the typical value for the latency is expected to be 10 ms on average [2]. Imperfect CSI has significant impacts on not only C-RAN but also wireless networks in general [20]. Since it is difficult to solve this problem using traditional information-theoretic approach [14], we take a stochastic optimization approach, which has well-developed mechanisms to address the impacts of noisy and delayed CSI. An optimal policy is found based on the particular structure of the topology configuration and rate allocation problem.

- *Response latency* experienced by cloud users has been recognized as one of the most important performance metrics in cloud computing [27,31]. Therefore, to improve MCC users' QoS, we model the response latency experienced by each MCC user as a constraint in our formulation.

- We investigate the trade-off between the systematic efficiency and the fairness among MCC users. A parameter is introduced to study such trade-off taking into account the delayed CSI in C-RAN. With this parameter, we re-formulate the problem to maximize the Jain's fairness index in MCC systems.

- Extensive simulations show that, with the emergence of MCC and C-RAN technologies, the design and operation of next-generation wireless networks can be significantly affected by cloud computing, and the proposed scheme is capable of achieving substantial performance gains over existing schemes.

The rest of the chapter is structured as follows. Section 8.2 describes the system. Section 8.3 discusses the issues caused by delayed CSI in C-RAN. We formulate the problem as a decision-theoretic problem in Section 8.4. Fairness and efficiency trade-off is studied in Section 8.5. Simulation results are discussed in Section 8.6. Finally, we conclude this study in Section 8.7 with future work.

8.2   System Description

In this section, we first describe MCC systems and dynamical configurations of C-RAN. Then, the physical layer and link-layer models for C-RAN are presented.

8.2.1   Mobile Cloud Computing with Cloud Radio Access Networks

The system we consider in this chapter is shown in Figure 8.1, which mainly consists of two sub-systems, that is, C-RAN and cloud computing. The problem addressed in this work crosses the two sub-systems. The wireless communication mainly happens at the C-RAN, while the processing (e.g., data mining) for the cloud services happens at the backend servers inside OTT service provider cloud.

In C-RAN, the traditional base stations are evolved into a system consisting of remote radio heads (RRHs) distributed in different geographic locations and a baseband processing unit (BBU) pool in the wireless cloud [9,19]. The RRHs are connected to the wireless network cloud via backhaul networks. Even optical networks can provide high-capacity and low-latency connections between RRHs and BBU pools, they are not always a feasible solution for backhaul networks. In contrast, microwave links and non-line-of-sight wireless links can provide more flexible deployments but introduce latency in the system [6]. Note that the implementation of the wireless network cloud varies. For example, in [41], the central processing and control unit is called "virtual base station pool." Nevertheless, we do not have any assumption on the implementation of the wireless network cloud. In the C-RAN considered in this chapter, there are B RRHs with one antenna each, which are denoted as a set $\mathcal{B}$.

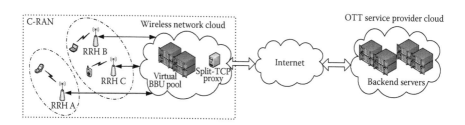

Figure 8.1   A cloud radio access network in the MCC environment. (From Cai, Y. et al., *IEEE Trans. Vehicular Technol.*, PP(99). With permission.)

Many mobile cloud services require the end-to-end reliable data transfer, TCP, across the two systems. There is a *split-TCP proxy* at the edge of the wireless network cloud. The split-TCP proxy is the split point for TCP flows. Such split-TCP proxy has been widely used in cloud computing [27] and traditional cellular networks [39]. In the context of cloud computing, split-TCP is also a popular scheme to provide reliable data transfer for cloud services. The client sets up a TCP connection to the nearby split-TCP proxy, then the split-TCP proxy sustains a persistent TCP connection to the data center with a very large TCP connection window [27]. In wireless networks, split-TCP proxy hides the wireless related issues from the wireline host via inserting a split point between the wireless and wired hosts. It locally acknowledges each segment and then stores and forwards the segments on the second TCP connection [39]. The split-TCP proxy can be implemented at system architecture evolution gateway (SAE-GW) in LTE systems [12] since the user data flows are tunneled to SAE-GW before being sent to the Internet.

Figure 8.2 shows the logical relationship of the mobile users, wireless network cloud, and OTT service provider cloud. TCP flows carrying mobile cloud services run from the mobile device to the backend server in OTT service provider cloud. Split-TCP proxy residing at the edge of wireless network cloud splits the end-to-end connection between the mobile user and the backend server into two connections

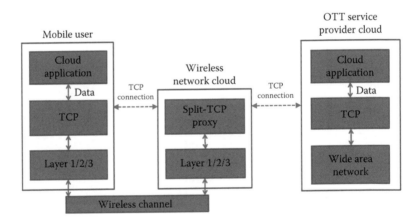

Figure 8.2 Logical protocol stacks of MCC system entities: mobile device, split-TCP proxy, and cloud backend server. (From Cai, Y. et al., *IEEE Trans. Vehicular Technol.*, PP(99). With permission.)

and sustains a persistent connection between itself and the back-end server. Meanwhile, the wireless network cloud conducts dynamic operations on wireless networks to provide best service for the upper layer. Such dynamic operations include topology configuration and rate allocation. Topology configuration controls how the RRHs cooperate with each other. For instance, in Figure 8.1, RRHs B and C form a cooperating set to serve the two MCC users together while RRH A itself is another cooperating set. After topology configuration, the wireless network cloud needs to decide the data rates that the MCC users can transmit. Inside a cooperating set, the signals are processed jointly such that there is no interference.

We denote the channel state matrix at time slot t as S^t, the topology configuration action at time slot t as Ω^t. The rate allocation vector has B elements, $\mathbf{R}^t = [R^{1,t} \cdots R^{B,t}]$, where each element is the rate allocation for a user. Here, we assume that there is only one user serviced by each RRH at a time slot, which is commonly assumed in the literature due to the opportunistic scheduling operation in practice [40]. The overall action is $a^t \triangleq \{\Omega^t, \mathbf{R}^t\} \in \mathcal{A}$, where $\mathcal{A}$ is the set of actions available. In this work, we assume that the power allocated to each user is the same, so that we can focus on the optimal topology configuration and rate allocation problem. On one hand, such binary power allocation has been shown to be optimal in some scenarios [13]. Thus it is reasonable to decouple power control at the time being. On the other hand, it is straightforward to adapt this framework to support power allocation by taking the possible power allocation scheme into the action space.

8.2.2   Physical Layer and Link Layer in C-RAN

In the following we introduce the physical layer and link-layer models. We attempt to make the modeling of these two layers to be as general as possible while sustaining a certain feasibility in performance analysis. This is because essentially C-RAN is supposed to be an open platform to support various technologies at lower layers [9].

Consider a cooperating set ω whose cardinality is $|\omega| = K$. Signals for mobile users served by RRHs in ω can be decoded without interfering with each other; while the mobile users served by the *non-cooperating* RRHs, $\mathcal{B} - \omega$, are interferers to ω. We number the RRHs in ω from 1 to K, and RRHs in $\mathcal{B} - \omega$ from $K + 1$ to B. Denote the complex channel gain from a mobile device served by RRH i to the antennas of all the RRHs in ω as $\mathbf{h}_i \in \mathbb{C}^{K \times 1}$, $i = 1, \dots, K, K + 1, \dots, B$. We assume that all mobile devices are allocated the same transmission power P, namely, the power allocation matrix of mobile devices in cooperating set ω is $\sqrt{P} \times \mathbf{I}_K$. If the complex data symbols of mobile devices served by cooperating set ω are $[x_1 \cdots x_K]$, and the data symbols of mobile devices served by the other RRHs are $[x_{K+1} \cdots x_B]$, the received signal of the antennas in a cooperating set ω is given by

$$\mathbf{y} = \sqrt{P} \sum_{l=1}^{K} \mathbf{h}_l x_l + \sqrt{P} \sum_{l'=K+1}^{B} \mathbf{h}_{l'} x_{l'} + \mathbf{n}, \tag{8.1}$$

where $\mathbf{n}$ is a vector of independent complex circularly symmetric additive Gaussian noise with each element $n \sim \mathcal{CN}(0, N_0)$. In the earlier signal level representation, the first term is the useful signal inside ω, while the second term is interference signal from $B - \omega$.

As a representative signal processing technique, minimum mean-square-error successive interference cancelation receiver [35, Ch. 10.1] can achieve the multiple access channel capacity. With fixed decoding order, the capacities of users ranging from $1, 2, \dots, K$ are given as follows:

$$C_K = \log\left(1 + \frac{P\|\mathbf{h}_K\|^2}{N_K}\right),$$

$$C_{K-1} = \log\left(1 + P\mathbf{h}_{K-1}^T (N_{K-1}I_K + P\mathbf{h}_K \mathbf{h}_K^*)^{-1} \mathbf{h}_{K-1}\right),$$

$$\cdots$$

$$C_1 = \log\left(1 + P\mathbf{h}_1^T \left(N_1 I_K + \sum_{l=2}^{K} P\mathbf{h}_l \mathbf{h}_l^*\right)^{-1} \mathbf{h}_1\right),$$

$$\tag{8.2}$$

where N_l, $l = 1, 2, \dots, K$ are the AWGN noise accounting for the receiver noise N_0 and the interference from outside ω. Specifically, the total noise at the lth antenna is $N_l = N_0 + P \sum_{l'=K+1}^{B} |\mathbf{h}_{l'}|^2$. Note that any other physical layer is likewise applicable to our work.

For a particular user u, if the current channel capacity is less than the transmission rate allocated, there is an outage so that the resulting transmission rate is 0; otherwise, the resulting transmission rate is equal to the allocated rate. In slot t, the performance of a user u is partially controlled by the action a^t taken by the wireless network cloud. The probability of error without any link-layer retransmission is defined as

$$p_{1,u} = \Pr\left(r_u^t(a^t) > C_u^t(a^t)\right),$$

(8.3)

where

r_u^t is the rate allocation for user u

C_u^t is the channel capacity at time slot t, which is a random variable since the channel state is unknown

For the link layer, we use hybrid automatic repeat request (HARQ) to reduce unreliability in wireless links. HARQ combines forward error correction and ARQ to increase the communication reliability. We assume a chase combining scheme in the following.* The performance of such an HARQ scheme has been analyzed in [22]. It is shown that, for user u, the number of packets transmitted, denoted as a random variable N_u, follows a Gaussian distribution with mean μ_u and variance σ_u^2

$$\mu_u = \frac{1 + p_{1,u} - p_{1,u}p_{2,u}}{1 - p_{1,u}p_{2,u}},$$

(8.4)

$$\sigma_u^2 = \frac{p_{1,u}(1 - p_{1,u} + p_{1,u}p_{2,u})}{1 - p_{1,u}p_{2,u}},$$

(8.5)

where

$p_{1,u}$ is the probability of error after decoding the information block by forward error correction

$p_{2,u}$ is the probability of error after soft combining the two successive transmissions of the same information block [22]

Note that $p_{1,u}$ is essentially the outage probability without HARQ defined in (8.3) and that $p_{2,u}$ is usually obtained via link level simulations [22].

Therefore, if the maximum number of transmissions allowed in the link layer is v, and if the action taken at time slot t is a^t, the packet error rate is

$$p_{e,u}(a^t) = \Pr(N_u > v) = Q\left(\frac{v - \mu_u}{\sigma_u}\right),$$

(8.6)

---

* Other HARQ schemes, such as incremental redundancy combining, are applicable in our work as well.

where $Q(\cdot)$ is the well-known Q-function. Moreover, the average transmission time of a TCP data packet over wireless links can be computed as

$$\overline{T}_{wireless,u}(a^t) = \mu_u \frac{L_{data} + L_{ack}}{r_u}, \tag{8.7}$$

where L_{data} and L_{ack} are the link-layer frame size for a TCP data packet and a TCP acknowledgment packet, respectively. We assume the downlink data rate of the C-RAN is the same as the uplink for ease of analysis.

8.3   C-RAN with Delayed CSI

In this section, we first introduce the channel modeling based on finite state Markov chains (FSMCs), which is essential in taking delayed CSI into account in our formulation. Then we discuss the delayed CSI issue in C-RAN followed by the belief-state concept that captures the uncertainty caused by delayed CSI.

8.3.1   Finite State Markov Chain Channel Model and Delayed CSI

We define the vector space consisting of B^2 elements as the system state $\mathcal{S}$. Assume at time slot t, the system state S^t is $s \in \mathcal{S}$, it will jump to s' at the next time slot. With the FSMC channel modeling [37], the state-transition function A is given by

$$A(s,s') = \Pr(S^{t+1} = s'|S^t = s) = \prod_{b=1,u=1}^{b=B,u=B} \Pr(I_{b,u}^{t+1}|I_{b,u}^t), \tag{8.8}$$

where $I_{b,u}^t$ and $I_{b,u}^{t+1}$ are the current state and next state of the FSMC from a transmit antenna of mobile user u to a receive antenna of RRH b.

To see how delay comes into C-RAN, we consider a C-RAN shown in Figure 8.1. The CSI is obtained via the pilot signals received at RRHs. After channel estimation, the CSI will be transmitted over backhaul networks to the wireless network cloud. At the wireless network cloud, a decision about how the RRHs cooperate and the rates at which MCC users can transmit are decided after obtaining CSI. Then, the user data are transmitted. Similar to the measurement and propagation of CSI, user signals are transmitted from MCC users to RRHs, and then are propagated over the backhaul networks. At the moment of decision making, the available CSI is delayed. We can abstract the total delay between the actual channel state at the moment of decision making and the one of observation as one single number. Then, we can map the delay in seconds into the transition steps in Markov chains. Therefore, the d-step transition probability is given by multiplying matrix A by d times, A^d.

8.3.2 Belief State with Known Delay Steps

Given the delay in steps, we can derive the *belief state*, which is the sufficient statistic of the previous action and observation history [17]. A belief state $\mathbf{b}^t$ at time slot t is a probability distribution of the state space. Accordingly, the probability that the state at time slot t being s^t is given by the corresponding element in $\mathbf{b}^t$, denoted as $b(s^t)$. Following [17], we use *belief state* to express both the vector $\mathbf{b}^t$ and its element given a state $b(s^t)$.

With techniques such as time-stamping, we can know the number of delay steps d. With such an assumption, the observation is just the actual state delayed by d steps. Denote the observation at time t as a random variable O^t. We have $O^t = S^{t-d}, t = d + 1,\ldots$ Thus we can derive the explicit relation between the current state and observation. The belief state is

$$
\begin{aligned}
b(s^{t+1}) &= \Pr(s^{t+1}|o^{t+1},o^t,\ldots) \\
&= \Pr(s^{t+1}|s^{t+1-d},s^{t-d},\ldots) \\
&= \Pr(s^{t+1}|s^{t+1-d}) \\
&= A^d(s^{t+1-d},s^{t+1}).
\end{aligned}
\tag{8.9}
$$

The third line is given by the first-order Markovian property assumed in the FSMC channel model, and A^d is the d-step probability transition matrix.

8.3.3 Belief State with Unknown Delay Steps

If it is difficult to get the number of delay steps, we can still compute the belief state based on Bayesian rule. We first introduce "Observation Function" $B(\cdot)$ [17]. Assume at time slot t, the observation of the system is $o \in \mathcal{O}.^*$ The observation function $B(\cdot)$ essentially depicts the probabilistic relationship between an observation $o \in \mathcal{O}$ and a

---

* Obviously, $\mathcal{O}$ is the same as $\mathcal{S}$. We use different notations for clear presentation.

state $s \in S$. Formally, observation is also a function of the action taken; however, in our problem here observation is independent of the action, so it is defined as

$$B(s,o) = \Pr(o|s). \tag{8.10}$$

State-transition function $A(\cdot)$ and observation function $B(\cdot)$ can be obtained via classical algorithms such as *Expectation Maximization* [29].

Essentially, provided a new observation at time $t+1$, o^{t+1}, the new belief should reflect the likelihood of ending up in new state s^{t+1}, the likelihood of observing o^{t+1}, and the pervious belief distribution $\mathbf{b}^t$. The rule to update the belief state according the previous belief and current observation based on the Bayesian rule [17] is

$$
\begin{aligned}
b(s^{t+1}) &= \Pr\left(s^{t+1}|o^{t+1}, \mathbf{b}^t\right) \\
&= \frac{B(s^{t+1}, o^{t+1}) \sum_{s^t \in S} A(s^t, s^{t+1}) b(s^t)}{\sum_{s^{t+1} \in S} B(s^{t+1}, o^{t+1}) \sum_{s^t \in S} A(s^t, s^{t+1}) b(s^t)}.
\end{aligned} \tag{8.11}
$$

With the belief state $\mathbf{b}^t$, we can compute the probability of error without link-layer retransmission Equation 8.3 as follows:

$$p_{1,u} = \sum_{c_u(s', a^t) < r_u^t} b(s'), \tag{8.12}$$

where $c_u(s', a^t)$ is the capacity that user u can achieve given the action a^t and the channel state s', and r_u^t is the rate allocation.

8.4   TCP Throughput over C-RAN in MCC Systems

In this section, we study the TCP throughput over C-RAN in MCC systems. Then we investigate the user response latency issue. Next, the TCP throughput maximization with response latency constraint problem is formulated as a constrained stochastic optimization problem. Finally, we derive the optimal topology configuration and rate allocation algorithm.

8.4.1   Round Trip Time and Split-TCP Throughput

Split-TCP is a popular reliable data transfer protocol for data center networks [27] and legacy cellular networks [39]. We can expect that it will play an important role in next-generation MCC systems.

Therefore, in this work, we adopt split-TCP as our transport layer protocol. A widely used TCP throughput model is developed in [25]. It has been used in cross-layer designs to maximize TCP throughput (for instance, [33]). In this section, we extend the existing work to take delayed CSI into account in the TCP throughput model.

We first discuss the round trip time (RTT). Figure 8.3 shows the RTTs for mobile cloud services over C-RAN. There are two types of RTTs [27]. RTT_1 represents the RTT between clients and the split-TCP proxy at the edge of wireless network cloud. RTT_2 is the RTT between the split-TCP proxy and the backend server in the OTT service provider cloud. We will not discuss the randomness in RTT_2 because this work is focused on the effect of C-RAN on cloud service.

RTT_1 consists of $T_{wireless}$ and $T_{backhaul}$, which represent the round trip transmission time over the wireless and backhaul networks, respectively. Note that, due to the wireless channel fading, $T_{wireless}$ is a random variable partly controlled by the action taken by the control unit in the wireless network cloud of C-RAN. The mean value of RTT_1 is given by

$$\overline{RTT}_1(a^t) = \overline{T}_{wireless}(a^t) + T_{backhaul},\qquad(8.13)$$

where $\overline{T}_{wireless}(a^t)$ is defined in Equation 8.7.

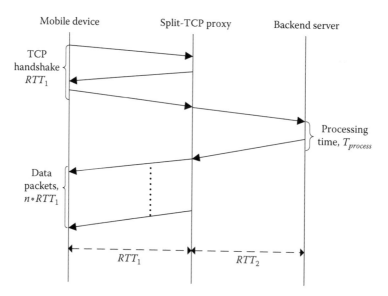

Figure 8.3 Round trip times and response latency using split-TCP. (From Cai, Y. et al., *IEEE Trans. Vehicular Technol.*, PP(99). With permission.)

We assume that the maximum segment size (MSS) is set in such a way that a single segment will fit into a single link-layer frame. That is, there is only one link-layer packet for a TCP segment. The queuing delay is not taken into consideration. Since we focus on the delayed CSI in this work, we can approximately take the queuing delay as a constant contribution to RTT. In terms of implementation, the queuing delay can be approximated by its statistical average. This assumption is reasonable because the time scale of queuing dynamics is generally larger than that of the wireless channel dynamics provided there is a bulk of data to transmit [10]. If the files are small, the queuing delay can be simply ignored. Therefore, we take queuing delay as a constant, which has been considered in $T_{backhaul}$.

Padhye et al. have developed a model for TCP connections. For user u, the average throughput can be derived as [25]

$$\bar{\eta} \approx \min\left\{ \frac{W_{max}}{\overline{RTT}}, \right.$$

$$\left. \frac{1}{\overline{RTT}\sqrt{2n_{ack}p_e/3} + T_0 \min\left\{1, 3\sqrt{3n_{ack}p_e/8}p_e(1 + 32p_e^2)\right\}} \right\},$$

(8.14)

where

W_{max} is the maximum congestion window

$\overline{RTT}$ is the RTT

n_{ack} is the number of packets acknowledged by a TCP ACK (generally 2)

T_0 is the initial time-out for the TCP sender

p_e is the TCP loss probability

The accuracy of such a model has been verified against real TCP traces in [25]. Note that the throughput of a TCP connection over a RAN, $\eta_{RAN,u}(a^t)$, is a random variable because the actual channel state S^t is unknown. $\bar{\eta}_{RAN,u}(a^t)$ is the mean value of it. For a connection in C-RAN, $\overline{RTT}$ and p_e are defined in Equations 8.13 and 8.6, respectively.

In split-TCP, the end-to-end throughput is the minimum throughput between the two TCP connections. For a user u, denote the average throughput of the TCP connection between the mobile user and split-TCP proxy as $\bar{\eta}_{RAN,u}$, and the one between split-TCP proxy and data centers as $\bar{\eta}_{cloud}$. The overall average throughput of split-TCP for u given the action taken a^t is

$$\bar{\eta}_u(a^t) = \min\left\{\bar{\eta}_{RAN,u}(a^t), \bar{\eta}_{cloud}\right\}. \tag{8.15}$$

8.4.2   Per-User Response Latency

Response latency experienced by cloud users is critical for mobile cloud services [27,31]. Since the main computation tasks are performed in data centers, MCC systems suffer from the response latency caused by processing time and communications among network entities. The processing latency is mainly caused by the hardware and the operating system, which is not the focus of our work. On the other hand, as will be shown in the following sections, the communication latency can be improved by careful design and operation of the C-RAN.

As shown in Figure 8.3, the connection between the client and split-TCP proxy spends about an RTT_1 in the handshaking phase. $T_{process}$ is the time needed for the backend servers to process the request. The split-TCP proxy needs to wait an RTT_2 and $T_{process}$ in setting up the connection and for the backend server to compute the results and to transmit them to the split-TCP proxy. Using the same assumption as [27], it takes $n*RTT_1$ to transmit the results from the split-TCP proxy to the MCC users. Denoting the total response latency as $\tau(a^t, S^t)$, we have

$$\tau(a^t, S^t) = (n+1)*RTT_1(a^t, S^t) + RTT_2 + T_{process}. \tag{8.16}$$

A typical value of n for search engine application is 4 [27]. Recall that RTT_1 at time slot t is a random variable depending on the actual system state S^t and the action taken a^t. Accordingly the average value of total response latency $\bar{\tau}$ is given by

$$\bar{\tau}(a^t) = (n+1)*\overline{RTT}_1(a^t) + \overline{RTT}_2 + T_{process}, \tag{8.17}$$

where
   $\overline{RTT}_1(a^t)$ is defined in Equation 8.13
   $\overline{RTT}_2$ and $T_{process}$ are considered to be constant

8.4.3 Maximizing TCP Throughput with Delayed CSI for Mobile Cloud Services

At time slot t, the system state S^t is an unobserved random variable. The wireless network cloud selects the cooperating RRHs and allocates the rate for MCC users, denoted as a^t. Denote the end-to-end throughput of a mobile user u given by Equation (8.15) as a random variable $\eta_u(a^t, S^t)$, then $\sum_{u=1}^{B} \eta_u(a^t, S^t)$ is the sum throughput of the system. The number of time slots considered is h, which is called the number of horizons in Markov decision process literature [17]. The cumulative rewards over h horizons is $\sum_{t=1}^{t=h} \sum_{u=1}^{u=B} \eta_u(a^t, S^t)$.

Accordingly, we denote the response latency defined in Equation 8.16 as τ_u, and we constrain the latency to be under a threshold α. To maximize the sum TCP throughput subject to the response latency constraint, we have the following optimization problem,

$$
\underset{a^t,\, t=1,2,\dots,h}{\text{maximize}} \quad \mathbb{E}\left[\frac{1}{h}\sum_{t=1}^{t=h}\sum_{u=1}^{u=B}\eta_u(a^t, S^t)\right] \tag{8.18}
$$
$$
\text{s.t.} \quad \mathbb{E}\left[\tau_u(a^t, S^t)\right] < \alpha, u = 1,\dots,B,\ \ t = 1,\dots,h.
$$

8.4.4 Greedy Policy

The problem in Equation 8.18 is a constrained stochastic optimization problem. We first propose a greedy policy, where the expected objective function value achievable at the current time slot is maximized. In other words, it is optimal for the stochastic optimization problem in Equation 8.18 when $h = 1$. Then, we will prove that the greedy policy is optimal when we consider multiple horizons. The procedures for the online and offline phases are also illustrated.

Theorem 8.1 *The optimal policy for the optimal topology configuration problem is given by Equation 8.19.*

$$
\underset{a^t}{maximize} \quad \mathbb{E}\left[\sum_{u=1}^{u=B}\eta_u(a^t, S^t)\right] \tag{8.19}
$$
$$
s.t. \quad \mathbb{E}\left[\tau_u(a^t, S^t)\right] < \alpha, u = 1,\dots,B.
$$

Proof: The optimality of the greed policy is proven via induction. Consider at horizon $h = 1$, the optimal action to take is the maximizer of $\mathbb{E}\left[\sum_{u=1}^{u=B}\eta_u(a^1)\right]$, which is obviously the action given by greedy policy to maximize the expected rewards in one step.

Assume at horizon $h, h \geq 1$, the optimal policy is the greedy policy given in Equation 8.19. Then at horizon $h+1$,

$$\frac{1}{h+1}\mathbb{E}\left[\sum_{t=0}^{t=h+1}\sum_{u=1}^{u=B}\eta_u(a^t,S^t)\right]$$

$$= \frac{1}{h+1}\mathbb{E}\left[\sum_{t=0}^{t=h}\sum_{u=1}^{u=B}\eta_u(a^t,S^t)\right] + \frac{1}{h+1}\mathbb{E}\left[\sum_{u=1}^{u=B}\eta_u(a^{h+1},S^{h+1})\right]. \quad (8.20)$$

So provided the hypothesis that the greedy policy maximizes the first term in the earlier equation, the action to take to maximize the total expected rewards is the one to maximize the second item, which is equivalent to the case with horizon 1. Therefore, the greedy policy is the optimal policy for problem in Equation 8.18.

From the channel observation and delay, we obtain the belief state, $\mathbf{b}^t$, which is the probability mass function of the current CSI. The stochastic optimization problem in Equation 8.19 can be converted into a deterministic optimization problem

$$\underset{a^t}{\text{maximize}} \quad \sum_{u=1}^{u=B}\overline{\eta}_u(a^t)$$

$$\text{s.t.} \quad \overline{\tau}_u(a^t) < \alpha, u = 1,\ldots, B. \quad (8.21)$$

Techniques to solve such integer programming problems have been well developed. For example, mediate size problems can be solved efficiently by the branch and bound method [18], and very large-scale integer programmings can be solved by heuristics such as Genetic Algorithm [23].

The algorithm to address the stochastic optimization Equation 8.18 includes an offline and online phases. In the offline phase, for each possible observation, a belief state is computed and the integer programming Equation 8.21 is solved. For the purpose of illustration, we adopt a brute-force approach in the offline phase, as shown in Algorithm 8.1. The optimal policy computed in the offline phase is utilized in the online phase to make the optimal decisions based on the delayed observations, as shown in Algorithm 8.2.

Algorithm 8.1 Brute-force search for the greedy policy: offline phase

---

1: {Find the feasible action at time slot t.}
2: **for** $o \in \mathcal{O}$ **do**
3:    Given the observation o, compute the belief state **b** according to Equation 8.9 or Equation 8.11.
4:    **for** $a \in \mathcal{A}$ **do**
5:       Compute the probability of error without any link-layer retransmission based on Equation 8.3, then the average transmission time over wireless channel and packet error rate for all users according to Equations 8.7 and 8.6, respectively.
6:       Calculate the expected throughput for split-TCP for cloud computing according to Equations 8.14 and 8.15, then store the value in a table TP_TABLE.
7:       Compute the expected response latency according to Equation 8.17, then store the value in a table L_TABLE.
8:       {The following step is used for the discussion on efficiency-fairness trade-off in Section 8.5.}
9:       Compute the Jain's index according to Equation 8.22, then store the value in a table $Jain\_TABLE$.
10:    **end for**
11:    {Brute-force search for the feasible action}
12:    **for** each response latency value τ in L_TABLE **do**
13:       **if** $\tau < \alpha$ **then**
14:          Store the corresponding action in a table FA_TABLE, they are the feasible actions.
15:       **end if**
16:       **if** FA_TABLE is not empty **then**
17:          Among the feasible actions in FA_TABLE, find the action achieving the maximum throughput in TP_TABLE, a^*. If there are more than one action achieving the same optimal throughput, an arbitrary one is taken to break the tie.
18:       **else**
19:          The optimization problem is infeasible, no action will be taken.
20:       **end if**
21:       Store a^* in the greedy policy table GP_TABLE
22:    **end for**
23: **end for**

Algorithm 8.2 Greedy policy: online phase

1: **for** $t = 0, 1, \ldots$ **do**
2:    Given the observation o^t, lookup the optimal action in table GP_TABLE. The action achieving the best objective function value is executed in the coming decision period.
3: **end for**

Discussion: In offline phase, the major computation cost lies on storage and manipulations of the station-transition function A. In the worst case, A is a $Q^{B^2} \times Q^{B^2}$ matrix, where Q is the channel quantization level. Fortunately, A is a sparse matrix with the majority of the non-zero elements along the diagonal. This hugely reduces the storage requirements and the manipulation complexity [28]. In the online phase, the complexity lies on selecting the best action achieving the best objective function value, which is equivalent to sorting an entry in GP_TABLE. So the complexity is $O(|\mathcal{A}|\log(|\mathcal{A}|))$ since the fundamental limit of run-time complexity for sorting algorithms in worst case is linearithmic [32]. In contrast, the complexity of computing the optimal policy for general partial observable Markov decision processes is PSPACE-complete, which is considered to be harder than NP-complete problems [8].

8.5 Fairness and Efficiency Trade-Off

As in many other multi-user systems, one important performance metric for MCC systems is fairness to the MCC users. This section investigates such metric based on the widely used Jain's fairness index [16]. For the system we consider, Jain's index is defined as

$$J(\overline{\eta}) = \frac{\left[\sum_{u=1}^{u=B} \overline{\eta}_u\right]^2}{\sum_{u=1}^{u=B} \overline{\eta}_u^2}, \tag{8.22}$$

where $\overline{\eta}$ is a vector, consisting of the throughput of all MCC users. Jain's index is between $1/B$ and 1 when the fairness to B users ranging from the least fair to the most fair.

We adopt the definition of optimal efficiency-fairness trade-off in [5]. In particular, we consider the action a^* obtaining the optimal efficiency-fairness trade-off, if there is not another action $a \neq a^*$ that satisfies either: (a) $\overline{\eta}(a) > \overline{\eta}(a^*)$ meanwhile $J(a) \geq J(a^*)$ or (b) $\overline{\eta}(a) \geq \overline{\eta}(a^*)$ meanwhile $J(a) > J(a^*)$. In practice, it is possible that even we achieve the optimal trade-off, the fairness index is still not satisfactory. Therefore, in such circumstance, we have to sacrifice some efficiency for the benefit of fairness.

To study the trade-off between the efficiency and fairness, we introduce a parameter β in the formulation, which is the percentage of throughput used to enhance the fairness in Jain's index. In particular, denote the maximum sum throughput we can have in Equation 8.21 as $\overline{\eta}^*$; we constrain the action such that the sum throughout is not less than $(1 - \beta) * \overline{\eta}^*$ while maximizing the Jain's index. Namely, we have the following problem,

$$\underset{a^t}{\text{maximize}} \quad J(a^t)$$

$$\text{s.t.} \quad \sum_{u=1}^{u=B} \overline{\eta}_u(a^t) \geq (1 - \beta) * \overline{\eta}^* \tag{8.23}$$

$$\overline{\tau}_u(a^t) < \alpha, u = 1, \ldots, B.$$

In the earlier optimization problem, the first constraint will be pushed to meet the right-hand side as close as possible, since with smaller efficiency, a higher extent of fairness can be expected. To solve this problem, the procedure in Algorithm 8.1 is modified to explore the trade-off. In particular, steps from 16 to 20 in Algorithm 8.1 are replaced by the sub-procedure defined in Algorithm 8.3.

Algorithm 8.3 Sub-procedure to explore the efficiency-fairness trade-off

---

1: **if** FA_TABLE is not empty **then**
2:      Among the feasible actions in FA_TABLE, find the maximum throughput in TP_TABLE, denoted as $\overline{\eta}^*$.
3: **else**
4:      Find the maximum throughput in TP_TABLE, denoted as $\overline{\eta}^*$.
5: **end if**
6: Identify the actions such that $\overline{\eta} > \overline{\eta}^* \cdot \beta$. Among them, find the one obtaining the maximum Jain fairness referring to $Jain\_TABLE$, a^*, which will be stored in GP_TABLE.

8.6 Simulation Results and Discussions

In this section, simulation results are presented to show the effectiveness of the proposed scheme. For the channel model and the physical system model are implemented in MATLAB®. The link-layer performance data is fed to network simulator, ns2 (version 2.34) [1]. We conduct simulations using the following settings. There are three RRHs in the C-RAN. The maximum size of a cooperating set is 2. The wireless channel is Rayleigh fading channel, and the normalized Doppler shift ranges from 0.01 to 0.06. The bandwidth is 45 KHz. The link layer allows frames to be transmitted at most three times. For TCP flows, the payload size is 760 bytes. W_{max} is 6 MSS. Other parameters are shown in Table 8.1. There are two existing schemes used for comparison. In the first one, the effects of imperfect CSI in C-RAN is not considered, and the topology configuration and rate allocation decisions are made based merely on current CSI observations to maximize TCP throughput in MCC systems, which is called "Existing scheme - perfect CSI." In the second one, TCP throughput in MCC systems is not considered, and the decisions are made to maximize the physical layer throughput based on imperfect CSI [7], which is called "Existing scheme–Physical layer throughput."

8.6.1 Performance Improvement

We measure CSI delay in C-RAN using the unit of samples, the same as in [15]. The performance metrics considered are sum TCP throughput of the MCC users and the average response latency among the

Table 8.1 Simulation Parameters

Parameter	Value
Carrier frequency	$2,110$ MHz
RRH antenna height	24 m
Mobile device antenna height	0.5 m
Number of multipath components	6
Sampling duration	$1/500,000$
Path loss	$30.18 + 26 * \log 10(distance)$
Receiver noise power density	-174 dBm/Hz
Mobile user transmit power	20 dBm
Per user latency constraint	350 ms

Source: Cai, Y. et al., *IEEE Trans. Vehicular Technol.*, PP(99). With permission.

MCC users. Figures 8.4 and 8.5 show the performance of the three schemes in the low mobility scenario where the normalized Doppler shift is 0.01. Figures 8.6 and 8.7 illustrate the results in the high mobility scenario where the normalized Doppler shift is 0.06. In the simulations, the response latency threshold α is set to be 0.35 s for the proposed scheme.

From these figures, we can observe that the proposed scheme outperforms the existing ones in terms of both system sum TCP throughput and response latency. In the low mobility scenario, the sum TCP throughput of both the proposed scheme and the existing scheme assuming perfect CSI in C-RAN drops slowly as the CSI delay increases. Nevertheless, the proposed scheme achieves more throughput than the existing scheme, for example, with the delay in CSI being 10 samples, by around 30%. Meanwhile, for the existing scheme assuming perfect CSI, the user response latency increases as the CSI gets more and more delayed. In the high mobility case, the proposed scheme can obtain higher throughput when the response latency is lower than the existing scheme assuming perfect CSI, as

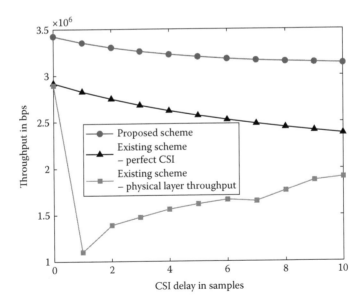

Figure 8.4   The effect of delayed CSI on the end-to-end TCP throughput in the low mobility case with normalized Doppler shift 0.01. (From Cai, Y. et al., *IEEE Trans. Vehicular Technol.*, PP(99). With permission.)

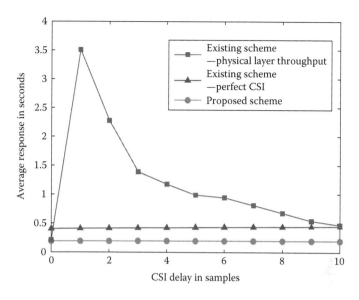

Figure 8.5 The effect of delayed CSI on the average response latency in the low mobility case with normalized Doppler shift 0.01. (From Cai, Y. et al., *IEEE Trans. Vehicular Technol.*, PP(99). With permission.)

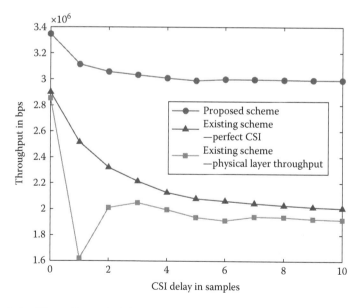

Figure 8.6 The effect of delayed CSI on the end-to-end TCP throughput in the high mobility case with normalized Doppler shift 0.06. (From Cai, Y. et al., *IEEE Trans. Vehicular Technol.*, PP(99). With permission.)

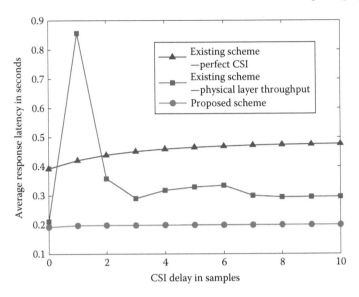

Figure 8.7   The effect of delayed CSI on the average response latency in the high mobility case with normalized Doppler shift 0.06. (From Cai, Y. et al., *IEEE Trans. Vehicular Technol.*, PP(99). With permission.)

shown in Figures 8.6 and 8.7. Note that when the delay is zero the proposed scheme still outperforms the existing schemes. That is because in the proposed scheme, we explicitly put the performance of split-TCP connections carrying MCC services into the formulation. In contrast, the existing schemes only aim to maximize physical layer throughput, which would not necessarily bring high performance for MCC.

In terms of throughput, Figures 8.4 and 8.6 show that the performance of the existing scheme only considering physical layer throughput is the worst among the three. These two figures indicate that the existing scheme maximizing the physical layer throughput does not guarantee a higher TCP throughput. In terms of response latency, Figures 8.5 and 8.7 show the results. In the low mobility case, as CSI delay increases, the response latency of the existing scheme maximizing physical layer throughput is getting close to that of the existing scheme assuming perfect CSI. In the high mobility case, it outperforms the existing scheme assuming perfect CSI when the CSI delay is larger than two samples. As shown in our previous work [7], the existing scheme maximizing the physical layer throughput has better performance than the existing scheme assuming perfect CSI when the criterion is the sum rates of all the MCC users in the system. Furthermore, its advantage decades as CSI delay increases. However, such a scheme is not appropriate when the criterion is the sum TCP throughput of mobile cloud services. The inherent reason is that the behavior of TCP is affected by not only the physical layer throughput but also the RTT and the end-to-end reliability. The existing scheme maximizing the physical layer throughput only strikes a balance between the outage probability and the rate allocation to achieve maximum physical layer throughout, which might be suboptimal for MCC systems. So, when the delay is small, for example, two samples, the existing scheme maximizing physical layer throughput has the worst performance. As the delay increases, the effectiveness of such a scheme in maximizing physical layer throughput decreases. Consequently, the TCP throughput and latency get close to the one under the existing scheme assuming perfect CSI. That is the reason why we can see a spike in the low CSI delay region in these figures.

Different from these two existing schemes, the proposed one not only considers the issue caused by the delayed CSI, more importantly, it also considers the ultimate performance of split-TCP carrying mobile cloud services. Hence, the simulation results indicate that the proposed scheme is the best one to dynamically configure the C-RAN in MCC systems. Therefore, we believe that it is critical to design and operate the wireless access network in the context of MCC, and the joint optimization can have significant advantages compared with the schemes where these two sub-systems are considered separately.

8.6.2   Effects of RTT_2 and $\bar{\eta}_{cloud}$

Figure 8.8 shows the effect of the RTT over wireline networks, RTT_2, on the response latency. In the figure, the response latency threshold is 0.35 s. With this threshold, feasible solutions can be found for Equation 8.18, such that the average response latency is bounded under the threshold. Note that if the response latency threshold is very small and/or RTT_2 is very large, it may not be possible to find feasible solutions to Equation 8.18. In this situation, other mechanisms, such as admission control, should be used to limit the number of MCC users in the system.

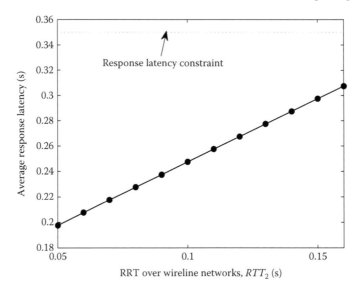

Figure 8.8 The effect of response latency threshold α in the high mobility case with normalized Doppler shift 0.06. (From Cai, Y. et al., *IEEE Trans. Vehicular Technol.*, PP(99). With permission.)

For an MCC user, the end-to-end throughput is decided by the two connections separated by the split-TCP proxy. The effect of the throughput of the connection between the split-TCP proxy and the OTT service provider cloud, $\overline{\eta}_{cloud}$, is shown in Figure 8.9. With the increase of $\overline{\eta}_{cloud}$, higher end-to-end throughput is observed. Moreover, the end-to-end throughput hits a plateau when $\overline{\eta}_{cloud}$ is large enough. That means the bottleneck lies in the RAN when the throughput between the split-TCP proxy and the OTT service provider cloud is sufficiently large. In this figure, we range the CSI delay from 0 to 8 samples. We can see that with the increase of CSI delay, the end-to-end throughput is getting smaller provided the same $\overline{\eta}_{cloud}$.

8.6.3 Efficiency-Fairness Trade-Off

An interesting observation in the simulation is the effect of β (the percentage of throughput defined in Section 8.5 to enhance the fairness in Jain's index) with different settings of CSI delay. Figures 8.10 and 8.11 show the results for the low mobility and high mobility scenarios,

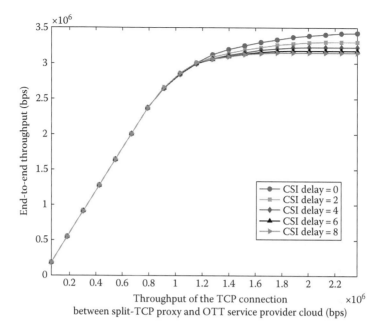

Figure 8.9 The effect of $\bar{\eta}_{cloud}$ in the low mobility case with normalized Doppler shift 0.01. (From Cai, Y. et al., *IEEE Trans. Vehicular Technol.*, PP(99). With permission.)

respectively. With the increase of β, more and more efficiency, that is, the system throughput, is traded-off for fairness. Yet, interestingly we observe that, with the increase of CSI delay, higher fairness is achieved provided the same setting of β. This implies that delay in CSI might help increasing the fairness even as we showed before that delayed CSI is bad for system throughput.

8.7 Conclusions

In this chapter, we jointly studied cloud-RAN and MCC in next-generation wireless networks. Particularly, the topology configuration and rate allocation problem in C-RAN has been investigated to improve the end-to-end TCP performance of MCC users in next-generation wireless networks. We proposed a decision-theoretic approach to tackle the imperfect CSI problem in C-RAN. The response latency experienced by each MCC user was modeled as a

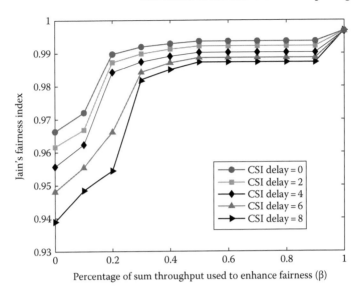

Figure 8.10 Efficiency-fairness trade-off in the low mobility case with normalized Doppler shift 0.01. (From Cai, Y. et al., *IEEE Trans. Vehicular Technol.*, PP(99). With permission.)

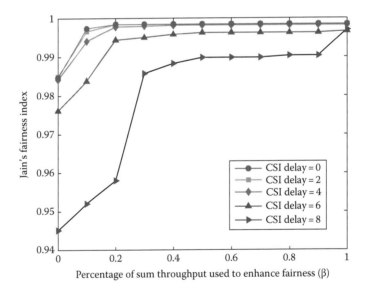

Figure 8.11 Efficiency-fairness trade-off in the high mobility case with normalized Doppler shift 0.06. (From Cai, Y. et al., *IEEE Trans. Vehicular Technol.*, PP(99). With permission.)

constraint. We also studied the trade-off between the efficiency and the fairness among MCC users. Using simulation results, we showed that our proposed scheme can significantly improve the system performance in terms of throughput and response latency of MCC users. In particular, the delayed CSI in C-RAN has significant effect on the performance, and our proposed scheme is able to reduce such effect, especially in large delay and high mobility scenarios.

References

1. Ns Network Simulator–Version 2. URL: http://www.isi.edu/nsnam/ns, accessed July 1, 2015.

2. 3rd Generation Partnership Project. Reply LS to R3-070527/R1-071242 on backhaul (X2 interface) delay. Technical report, 3rd Generation Partnership Project, 2007.

3. H. Balakrishnan, V. N. Padmanabhan, S. Seshan, and R. H. Katz. A comparison of mechanisms for improving TCP performance over wireless links. *IEEE/ACM Trans. Netw.*, 5(6):756–769, Dec. 1997.

4. S. Bhaumik, S. P. Chandrabose, M. K. Jataprolu, G. Kumar, A. Muralidhar, P. Polakos, V. Srinivasan, and T. Woo. CloudIQ: A framework for processing base stations in a data center. In *Proceedings of ACM Mobicom'12*, Istanbul, Turkey, 2012.

5. A. Bin Sediq, R. H. Gohary, R. Schoenen, and H. Yanikomeroglu. Optimal tradeoff between sum-rate efficiency and Jain's fairness index in resource allocation. *IEEE Trans. Wireless Commun.*, 12(7):3496–3509, 2013.

6. D. Bojic and NEC Europe. Advanced wireless and optical technologies for small-cell mobile backhaul with dynamic software-defined management. *IEEE Commun. Mag.*, 51(9):86–93, 2013.

7. Y. Cai, F. R. Yu, and G. Senarath. Optimal clustering and rate allocation for uplink coordinated multi-point (CoMP) systems with delayed channel state information (CSI). In *Proc. IEEE ICC'13*, Budapest, Hungary, June 2013.

8. A. R. Cassandra. Exact and approximate algorithms for partially observable Markov decision processes. Brown University, Providence, RI, 1998.

9. China Mobile Research Institute. C-RAN: The road towards green RAN. Technical report. http://labs.chinamobile.com/, accessed: July 18, 2013.

10. Y. Cui, Q. Huang, and V. K. N. Lau. Queue-aware dynamic clustering and allocation for network MIMO systems via distributed stochastic learning. *IEEE Trans. Signal Proc.*, 59(3):1229–1238, Mar. 2011.

11. H. T. Dinh, C. Lee, D. Niyato, and P. Wang. A survey of mobile cloud computing: Architecture, applications, and approaches. *Wireless Commun. Mobile Comput.*, 13(18):1587–1611, Dec. 2013.

12. V. Farkas, B. Hder, and S. Novczki. A split connection tcp proxy in LTE networks. In R. Szab and A. Vidcs, eds., *Information and Communication Technologies*, vol. 7479 of Lecture Notes in Computer Science, pp. 263–274. Springer Berlin, Heidelberg, Germany, 2012.

13. A. Gjendemsj, D. Gesbert, G. E. Oien, and S. G. Kiani. Binary power control for sum rate maximization over multiple interfering links. *IEEE Trans. Wireless Commun.*, 7(8):3164–3173, 2008.

14. A. Goldsmith, M. Effros, R. Koetter, M. Medard, A. Ozdaglar, and L. Zheng. Beyond Shannon: The quest for fundamental performance limits of wireless ad hoc networks. *IEEE Commun. Mag.*, 49(5):195–205, May 2011.

15. K. Huang. MIMO networking with imperfect channel state information. PhD thesis, The University of Texas at Austin, Texas, 2007.

16. R. Jain, D.-M. Chiu, and W. R. Hawe. A quantitative measure of fairness and discrimination for resource allocation in shared computer system. Eastern Research Lab., DEC, 1984.

17. L. P. Kaelbling, M. L. Littman, and A. R. Cassandra. Planning and acting in partially observable stochastic domains. *Artificial Intelligence*, 101(1-2):99–134, May 1998.

18. A. H. Land and A. G. Doig. An automatic method of solving discrete programming problems. *Econometrica*, 28(3):497–520, Jul. 1960.

19. Y. Lin, L. Shao, Z. Zhu, Q. Wang, and R. K. Sabhikhi. Wireless network cloud: Architecture and system requirements. *IBM Journal of Research and Development*, 54(1):4:1–4:12, 2010.

20. D. J. Love, R. W. Heath, V. K. N. Lau, D. Gesbert, B. D. Rao, and M. Andrews. An overview of limited feedback in wireless communication systems. *IEEE J. Sel. Areas Commun.*, 26(8):1341–1365, Oct. 2008.

21. C. Luo, F. R. Yu, H. Ji, and V. C. M. Leung. Cross-layer design for TCP performance improvement in cognitive radio networks. *IEEE Trans. Veh. Tech.*, 59(5):2485–2495, Jun. 2010.

22. M. Assaad and D. Zeghlache. Comparison between MIMO techniques in UMTS-HSDPA system. In *Proceedings of IEEE Eighth International Symposium on Spread Spectrum Techniques and Applications*, Sydney, Australia, pp. 874–878, 2004.

23. M. Mitchell. *An Introduction to Genetic Algorithms*. A Bradford Book, 3rd edition, Feb. 1998.

24. NIST. The NIST definition of cloud computing (draft). Technical Report Special Publication 800-145 (Draft), Jan. 2011.

25. J. Padhye, V. Firoiu, D. F. Towsley, and J. F. Kurose. Modeling TCP Reno performance: A simple model and its empirical validation. *IEEE/ACM Trans. Netw.*, 8(2):133–145, Apr. 2000.

26. G. Pallis. Cloud computing: The new frontier of internet computing. *IEEE Internet Comput.*, 14(5):70–73, 2010.

27. A. Pathak, Y. A. Wang, C. Huang, A. Greenberg, Y. Charlie Hu, R. Kern, J. Li, and K. W. Ross. Measuring and evaluating TCP splitting for cloud services. In *Proceedings of the 11th International Conference on Passive and Active Measurement (PAM'10)*, pp. 41–50, Berlin, Germany, 2010. Springer-Verlag.

28. S. Pissanetzky. *Sparse Matrix Technology*. Academic Press, 1984.

29. L. R. Rabiner. A tutorial on hidden Markov models and selected applications in speech recognition. *Proc. IEEE*, 77(2):257–286, Feb. 1989.

30. O. S.-H. Park, O. Simeone, O. Sahin, and S. Shamai. Robust and efficient distributed compression for cloud radio access networks. *IEEE Trans. Veh. Tech.*, 62(2):692–703, Feb. 2013.

31. M. Satyanarayanan, P. Bahl, R. Caceres, and N. Davies. The case for VM-based cloudlets in mobile computing. *IEEE Pervasive Comput.*, 8(4):14–23, 2009.

32. M. Sipser. *Introduction to the Theory of Computation*. Cengage Learning, 2006.

33. A. L. Toledo, X. Wang, and B. Lu. A cross-layer TCP modelling framework for MIMO wireless systems. *IEEE Trans. Wireless Commun.*, 5(4):920–929, Apr. 2006.

34. N. H. Tran, C. S. Hong, and S. Lee. Cross-layer design of congestion control and power control in fast-fading wireless networks. *IEEE Trans. Parallel Dist. Syst.*, 24(2):260–274, Feb. 2013.

35. D. Tse and P. Viswanath. *Fundamentals of Wireless Communication*. Cambridge University Press, 2005.

36. G. Wang and T. S. E. Ng. The impact of virtualization on network performance of Amazon EC2 data center. In *Proceedings of IEEE INFOCOM'10*, San Diego, CA, Mar. 2010.

37. H. S. Wang and N. Moayeri. Finite-state Markov channel. A useful model for radio communication channels. *IEEE Trans. Veh. Tech.*, 44(1):163–171, Feb. 1995.

38. S. Wang and S. Dey. Adaptive mobile cloud computing to enable rich mobile multimedia applications. *IEEE Trans. Multimedia*, 15(4):870–883, 2013.

39. W. Wei, C. Zhang, H. Zang, J. Kurose, and D. Towsley. Inference and evaluation of split-connection approaches in cellular data networks. In *Proceedings of Passive and Active Measurement Conference*, Adelaide, Australia, 2006.

40. R. Xie, F. R. Yu, H. Ji, and Yi Li. Energy-efficient resource allocation for heterogeneous cognitive radio networks with femtocells. *IEEE Trans. Wireless Commun.*, 11(11):3910 –3920, Nov. 2012.

41. Z. B. Zhu, P. Gupta, Q. Wang, S. Kalyanaraman, Y. Lin, H. Franke, and S. Sarangi. Virtual base station pool: Towards a wireless network cloud for radio access networks. In *Proceedings of the Eighth ACM International Conference on Computing Frontiers*, New York 2011.

CHAPTER 9

Cloud-RAN-Based Small Cell Networks with Wireless Virtualization

Heli Zhang, Weidong Wang, Xi Li, and Hong Ji

CONTENTS

9.1   Introduction

In recent years, with the fast development of information and communications technology (ICT), mobile access through smart phones and iPads has been increasing highly. The Cisco report of 2015 predicts that the worldwide mobile traffic will reach 24.3 EB by the year 2019 [1]. Cloud radio access network (C-RAN) based small cell architecture has attracted much attention due to its availability of sustaining a higher traffic volume [2]. Researchers believed that this architecture can play a key role in future 5G mobile networks. Wireless network virtualization [3], with the advantage of enabling abstraction and sharing of infrastructure and radio spectrum resources, reducing capital expenditure (CAPEX) and operational expenditure (OPEX) of wireless (radio) access networks as well as core networks, can be applied to enhance the C-RAN architecture and is becoming a promising research direction [4].

Large amounts of works have been done for wireless network virtualization. To satisfy the requirement of efficient resource utilization of virtualization, authors in [5] proposed a resource negotiation based solution, which is applied to Long-Term Evolution-Advanced (LTE-A) environments with numerous small cells. In this solution, wireless resources are mapped to radio virtualization elements and are allocated reasonably to achieve higher throughput. With the same objective, authors of [6] investigated the virtualization of wireless resource in LTE systems, where a shared RAN connected to multiple mobile network operators (MNOs) and users was put forward. Kamel et al. [7] also paid much attention to the virtualization of wireless access networks and proposed a dynamic resource allocation scheme for the schedule of radio resource blocks, and meanwhile, the fairness between cell-edge users and cell-center users are considered. Different from previous works focusing on only one physical wireless network, authors of [8] studied the resource utilization problem under multiple physical networks. An opportunistic spectrum sharing method was put forward to increase spectrum allocation efficiency. Besides of efficient resource utilization, isolation is also important to virtualization. In view of this, D. Garlisi et al. [9] presented a virtualization solution for wireless local area networks. With this scheme, flexible resource partitioning is achieved based on the concept of MAClets. In [10], authors separated the infrastructure providers (InPs) and service providers (SPs) in a virtualized framework; the interaction between InPs and SPs was modeled by the stochastic game, and the prices charged by InPs and SPs are determined while taking the quality-of-service (QoS) of the users into account. To meet mobile requirements of future network, Hoffmann et al. [11] concerned three elements that are virtualized physical resources, virtual resource manager, and virtual network controller, respectively. Furthermore, one service-specific algorithm is showed to adapt the variation of mobile networks. In [12], one converged infrastructure of next generation that supporting cloud and mobile cloud computing services was presented, where the virtualization of heterogeneous networks and related challenges were discussed.

Although some excellent works have been done on wireless network virtualization, when combing it with C-RAN based small cell network (SCN), most works pay attention to the construction of architecture, the virtualization and allocation of wireless resources, less focuses on the design of user association technology. User association that determines appropriate small cells for users has many advantages such as balancing the load, etc. Since the small cells are densely deployed and the mobile users' positions change rapidly, disassociation happens frequently. Therefore, investigating user association scheme is necessary in C-RAN SCN. In this chapter, we first construct a user association optimization problem in C-RAN SCN with mobile users. When mobility is considered in the network, one major limitation is the latency experienced by users to reach the cloud center. Users are acutely sensitive to delay: as latency increases, interactive response suffers. Since the interaction time foreseen in 5G systems are quite small (in the order of milliseconds) [13], a strict latency control must be somehow incorporated. Thus, the aim of the objective function in the user association optimization problem is to constrain the overall network latency. The function of latency is deduced from user's transmission rate. Moreover, since the increasing CO_2 emissions by ICT can damage the worldwide environment and the interference generated by neighboring base stations can lower the overall throughput of the network, energy saving and interference limitation are also incorporated in the optimization problem and are regarded as two important targets. In the C-RAN based SCN with wireless virtualization, small cells are managed by the InPs and users are served by the SPs. To achieve the target of efficient resource utilization, we assume that users are free to access the small cells when appropriate prices are charged.

We summarize our contributions in the following:

1. Formally define a user association optimization problem and present a formulation that combines the network latency and prices charged by SPs for users' free access to small cells.

2. Energy saving and interference limitation are incorporated as two objectives when establishing the optimization problem.

3. Present a three-phase search algorithm (TPSA) based on the rule of multicriterion multimodal assignment problem (MMAP) to solve the proposed problem and search the appropriate small cells for users.

The rest of the chapter is organized as follows. Section 9.2 provides a description of the user association optimization problem and the system model. The problem is analyzed and in Section 9.3, moreover, the proposed three-phase search approach is also discussed. The numerical results and their discussion are presented in Section 9.4 followed by conclusions in Section 9.5.

9.2 System Model

9.2.1 Network Model

The C-RAN based small cell architecture with wireless virtualization [14] is shown in Figure 9.1, which includes the cloud part and the access part. The cloud part is comprised of baseband process unit (BBU) pools implementing most of the computation algorithms such as digital baseband processing, mobility management, resource allocation, user association, etc. The access part is formed by small cells, which are managed by two different InPs, called InP1 and InP2, respectively. Users are also belonged to different SPs, and SPs are responsible of providing rich services to users. Different from

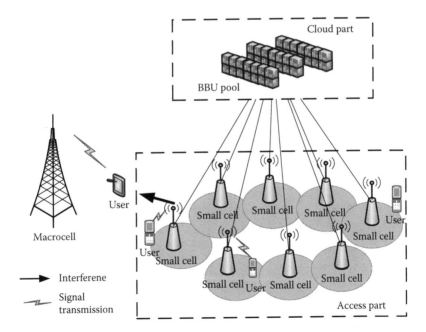

Figure 9.1 Network architecture.

traditional network that users only be allowed to access to the small cell under the control of one legitimate operation, in this architecture, under the concept of wireless virtualization, users are free to choose the appropriate small cell to connect when some money is paid to the InP. The price described in Section 1.2.3.2 for access is determined by how much utility is obtained by the user and SP. Small cells are usually deployed overlay a macrocell. To avoid the interference generated from small cells to the macrocell, the physical channels occupied by the macrocell should be unavailable or used under a power constraint for small cells. Thus, under this interference avoidance rule, the cloud part is responsible of determining the available physical channels for small cells.

The small cell vector is defined as $S = \{1,\ldots,S\}$ and let the user set be U. Assume the vector of orthogonal channels for small cell n be $L_n = \{1,\ldots,L_n\}, n \in S$, where the width of the each is ω. In order of limiting interference to the macrocell, the maximum power transmitted on the corresponding channel is constrained. Then, the power vector of nth small cell provided by the cloud part can be $p_n = \{p_{n1},\ldots,p_{nl},\ldots,p_{nL_n}\}$. p_{nl} means that the power on the lth block of the nth small cell. This symbol also signifies that once the physical channel is allocated to the user, the transmit power is determined. $T_{i,l}^n$ is the user association index, if at time t user i access to small cell n, the value of the index is designated to 1, otherwise is 0.

9.2.2 Physical Link Model

Assume the channel state information can be obtained through channel estimation in BBU, the SINR of user i in small cell n utilizing the channel l can be written as

$$\gamma_{i,l}^n(\tau) = \frac{h_{i,l}^n(\tau)p_{i,l}^n(\tau)}{\sum_{k \neq n, k \in S} g_{i,l}^k(\tau)p_{i,l}^k(\tau) + I_l + N_0}$$

$$p_{i,l}^n(\tau), p_{i,l}^k(\tau) \in \mathbf{p_n} \tag{9.1}$$

where $h_{i,l}^n(\tau)$ and $p_{i,l}^n(\tau)$ denote the channel gain and power between the small cell n and user i. $g_{i,l}^k(\tau)p_{i,l}^k(\tau)$ denotes the interference from neighboring small cell. $g_{i,l}^k(\tau)$ is the channel gain between the neighboring small cell k and the user i located at small cell n. $p_{i,l}^k(\tau)$ is the power generated by small cell. N_o denotes the white Gaussian noise. I_l is the inter-channel interference that happens due to users' mobility that $I_l = (f_d)^2/2 \sum_{j=1, j\neq l}^{L_n} 1/j - l$, $f_d = vf_c/c$. v and f_c are users' travel speed and channel central frequency.

Assume $v_{i,l}^n(\tau)$ be the transmission rate gained by user i at time $[t-1, t]$, then the following function can be obtained

$$v_{i,l}^n(\tau) = w \log(1 + \gamma_{i,l}^n(\tau)) \text{bps/Hz} \qquad (9.2)$$

9.2.3   Utility Definition

Utilities of small cells play an important role in determining whether one user is allowed to associate with the small cell. Through carefully designing the utility function, we intend to reach that when one users' association request decreases the utility of the small cell, then the user will be refused to access. Here, the utility is related to two elements, see the network latency and the price charged by the InP. In the following, we describe the two elements in detail.

9.2.3.1   Network Latency

Latency in this chapter refers to the time needed to transmit unit bit on the bandwith of w. The minimal potential delay fairness function [15] with the advantage of guaranteeing the network throughput and users' fairness is applied to deduce the network latency. This function can be denoted as the reciprocal function of user's transmission rate, and with which the latency for one certain user can be obtained. In the following, we show the latency of user i:

$$\frac{1}{v_{i,l}^n(\tau)} = \frac{1}{w \log(1 + \gamma_{i,l}^n(\tau))} \qquad (9.3)$$

The network latency is an assembly of all users' latency, which can be written as

$$\sum_{n \in S} \sum_{i \in U} T_{i,l}^n(\tau) \frac{1}{v_{i,l}^n(\tau)} \qquad (9.4)$$

9.2.3.2 Access Price

As described in the network description, the users should pay the money if they want to access to the small cell without a license. The price is determined by the SP, and it should be a constant in this chapter, we define the mathematical symbol as δ.

With network latency and access price, the utility function is formulated by the weighted sum approach. Thus, in the following we give the network utility represented by the U^n for small cell n:

$$U^n = \sum_{n \in S} \sum_{i \in U} \left(\alpha T_{i,l}^n(\tau) \frac{1}{v_{i,l}^n(\tau)} + \beta \delta \right) \tag{9.5}$$

where α and β are two parameters to adjust the order of magnitude of the latency and price. From the earlier function, it's obvious that the utility will decrease when the network latency increases, then the user will not be permitted to the cell, and vice versa. For the access price, it has the same influence on the user association.

9.2.4 Optimization Problem

In this section, we formulate the user association problem while guaranteeing the network utility. Furthermore, considering the energy and interference problems, energy saving and interference limitation are also regarded as two objectives in the user association optimization energy saving a problem. With the two targets, the energy consumed by the network and the interference suffered from other small cells should be minimized, which are shown in the following:

$$\min \sum_{n \in S} \sum_{i \in U} \sum_{l \in L_n} p_{i,l}^n(\tau) T_{i,l}^n(\tau)$$

$$\min \sum_{m \in S} \sum_{i \in U} \sum_{l \in L_n} g_{i,l}^m(\tau) p_{i,l}^m(\tau) T_{i,l}^n \tag{9.6}$$

where

$g_{i,l}^m(\tau) p_{i,l}^m$ is the interference generated from the mth small cell to the ith user located at the nth small cell

$g_{i,l}^m$ and $p_{i,l}^m$ are the channel gain and transmitting power from the mth small cell to user i

Combining with the network utility, the optimization problem can be formulated as the following:

$$\min \; U$$

$$s.t. \; \min \sum_{n \in S} \sum_{i \in U} \sum_{l \in L_n} p_{i,l}^n(\tau) T_{i,l}^n(\tau)$$

$$\min \sum_{m \in S} \sum_{i \in U} \sum_{l \in L_n} g_{i,l}^m(\tau) p_{i,l}^m(\tau) T_{i,l}^n$$

$$C.1 \; \sum_{n \in S} \sum_{l \in L_n} T_{i,l}^n(\tau) = 1$$

$$C.2 \; T_{i,l}^n(\tau) \in \{0,1\} \tag{9.7}$$

Where function $C.1$ means at time τ every user can only access to one small cell. Within this optimization problem, the optimal small cells are selected out for users with lower network latency, less energy consumption, and interference influence.

9.3   The Three-Phase Search Algorithm

9.3.1   Problem Analysis

The user association optimization problem shown in Equation 9.7 is equal to a multicriterion MMAP [16], where multiple objectives should be achieved. To solve the MMAP, weighted sum method and delamination sequence method are commonly utilized. The former merges the multiple objectives to one and the user–small cell pair can be figured out with Newton algorithm. The later transform the MMAP to several sub-problems, where each has one or two objectives, and then the user–small cell pair is found out through solving the sub-problems in sequence. In this chapter, based on the work of Pedersen et al. [16], we utilize the delamination sequence method and propose a TPSA.

The user association optimization problem is divided into two sub-problems, which are bicriterion multimodal assignment problem (BiMMP, see function Equation 9.8) and optimal search sub-problem (OSP, see function in Equation 9.9), respectively.

$$\min \sum_{n\in S}\sum_{i\in U}\sum_{l\in L_n} p_{i,l}^n(\tau)T_{i,l}^n(\tau)$$

$$\min \sum_{m\in S}\sum_{i\in U}\sum_{l\in L_n} g_{i,l}^m(\tau)p_{i,l}^m(\tau)T_{i,l}^n \qquad (9.8)$$

$$C.1 \quad \sum_{n\in S}\sum_{l\in L_n} T_{i,l}^n(\tau) = 1$$

$$C.2 \quad T_{i,l}^n(\tau) \in \{0,1\}$$

$$\min U$$

$$s.t. \ T_{i,l}^n \in \mathcal{T} \qquad (9.9)$$

Different from the single objective optimization, the BiMMP should satisfy multiple criterions. Then the feasible solution is not unique. Thus, through solving the problem of BiMMAP, a group of feasible solutions denoted by $\mathcal{T}$ can be found. By solving the problem of Equation 9.9, the optimal solution is selected from $\mathcal{T}$. Based on this optimal solution, whether the user is allowed to access to the small cell is determined.

9.3.2 Pareto Definition

For single-criterion optimization, the concept of optimality is well defined. Respecting common practice in the field of multi-criterion optimization, the Pareto concept of optimality is deployed [17]. The solutions in $\mathcal{T}$ should meet the Pareto Optimization. In the following, we will define the Pareto Optimization. Assume that

$$f_1 = \sum_{n\in S}\sum_{i\in U}\sum_{l\in L_n} p_{i,l}^n(\tau)T_{i,l}^n(\tau) = \mathbf{PT}$$

$$f_2 = \sum_{m\in S}\sum_{i\in U}\sum_{l\in L_n} g_{i,l}^m(\tau)p_{i,l}^m(\tau)T_{i,l}^n = \mathbf{\Gamma T} \qquad (9.10)$$

Then one point can be constructed as $f = \{f_1, f_2\}$. One solution can generate one corresponding point according to function in Equation 9.10, searching for the points that satisfy optimization function in Equation 9.7 is equal to find the feasible solutions.

Definition 9.1 *Pareto solution, if the feasible solutions* $\mathbf{T}^1, \mathbf{T}^2 \in \mathcal{T}$ *of Equation 9.7 have the following relationship,*

$$f^1 \le f^2 \Leftrightarrow f_r^1 \le f_r^2, r \in \{1,2\} \qquad (9.11)$$

Then we call that point f^2 is dominated by f^1. If the point cannot be dominated by any point, the corresponding solution is a Pareto solution.

Definition 9.2 When the following constraint is satisfied, $\mathcal{T}_\mathcal{E}$ is the effective solution vector for Equation 9.8

$$\mathcal{T}_\mathcal{E} = \{\mathbf{T} \in \mathcal{T} | \text{No solution } \tilde{\mathbf{T}} \text{ exists and}$$

$$\text{satisfies that } \tilde{\mathbf{T}} \in \mathcal{T} : (\mathbf{P}\tilde{\mathbf{T}}, \mathbf{\Gamma}\tilde{\mathbf{T}}) \text{dominate}(\mathbf{PT}, \mathbf{\Gamma T})\} \tag{9.12}$$

Definition 9.3 Based on the inequality of Equation 9.11, $f_\mathbb{N}$ can be a vector of non-dominated point.

$$f_\mathbb{N} = \{(f_1, f_2) \in \mathbb{Z}^2 | f^1 = \mathbf{P}\tilde{\mathbf{T}}, f^2 = \mathbf{\Gamma}\tilde{\mathbf{T}}, \tilde{\mathbf{T}} \in \mathcal{T}_\mathcal{E}\} \tag{9.13}$$

$f_\mathbb{N}$ can be partitioned into supported points and unsupported points. Supported points can be further subdivided into extreme and non-extreme.

As shown in Figure 9.2, assume the area constructed by the lines covers the whole point space. Then the points on the edge can be called supported nondominated points, such as $f^1 \sim f^5$. When these points stay on the vertex of the point space, they are called extreme points, such as $f^1 \sim f^3$ and f^5. In addition, in the triangle area constructed by the supported and non-dominated points, we can find the unsupported points, for example, f^6 is the one unsupported point occurs in the triangle area built by f^2 and f^3.

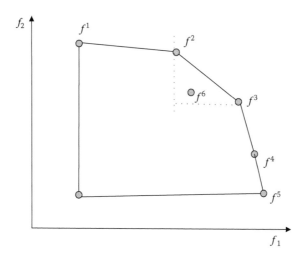

Figure 9.2   Point space.

9.3.3 The Three-Phase Search Algorithm

The two-phase approach presented in [17] is efficient for solving the BiMMAP. Combing the phase for solving the OSP, a TPSA is introduced to search the user association solution for the network. Just as the name implies, the algorithm includes three stages: first, in phase I find the supported non-dominated points and record the corresponding user association solutions (Algorithms 9.1 and 9.2). Second, in phase II, searching the supported non-extreme and the unsupported non-dominated points with the corresponding user

---

Algorithm 9.1 Search unsupported non-dominated points

1: **Initialization:**

    a) $f^{UL} := (\mathbf{PT^{UL}}, \Gamma\mathbf{T}^{UL})$, where $\mathbf{T^{UL}}$ is the optimal solution of $lex\ min = (\mathbf{PT}, \Gamma\mathbf{T})$.

    b) $f^{DR} := (\mathbf{PT^{DR}}, \Gamma\mathbf{T}^{DR})$, where $\mathbf{T^{DR}}$ is the optimal solution of $lex\ min = (\mathbf{PT}, \Gamma\mathbf{T})$.

    c) Assume initial solution space be $\mathcal{T} = \{\mathbf{T^{UL}}, \mathbf{T^{DR}}\}$.

2: **Search Phase:**

    a) If $f^{UL} = f^{DR}$, then the algorithm stops

Only one non-dominated solution exists in the solution space).

    b) $\mathcal{F} := (f^{UL} = f^{DR})$

    c) $f^+ := f^{UL}, f^- := f^{DR}$

    d) While $f^+ \neq f^-$, do {

      $\lambda = \lambda(f^+, f^-)$

      Obtain the optimal solution $\mathbf{T^*}$ from Equation 9.13 and $f^*_\lambda = (f^*_1, f^*_2)$

      If $f^*_\lambda < (\lambda f^+_1 + f^+_2)$, then

      {

         $\mathcal{F} = \mathcal{F} \bigcup \{f^*\}$

         $\mathcal{T} = \mathcal{T} \bigcup \{\mathbf{T^*}\}$

      }

      Else{

         $f^+ = f^-$,

         $f^- = \text{NEXT}(\mathcal{F}, f^+)$

      }

    }

    End While

   END

Algorithm 9.2 Search the supported non-extreme and the unsupported non-dominated points

---

1: Input parameter and construct the triangle $\Delta(f^+, f^-)$
2: $\lambda = \lambda(f^+, f^-)$,
   $\mathcal{F} := \{f^+, f^-\}$
3: $t = 1$,
   $LB := \lambda f_1^+ + f_2^+$,
   $UB := updateUB\mathcal{F}$
4: While $LB \leq UB$, do
   $\quad\quad f^t, \mathbf{T^t} := \mathbf{KBest}(t, \lambda)$
   $\quad\quad$ IF $NonDom(f^t)$, then
   $\quad\quad\quad \mathcal{F} := \mathcal{F} \cup f^t$,
   $\quad \mathcal{T} = \mathcal{T} \cup \mathbf{T^t}$
   $\quad\quad\quad UB := updateUB(\mathcal{F})$
   $\quad\quad$ END IF
   $\quad\quad LB := \lambda f_1^t + f_2^t$,
   $\quad t = t + 1$

   END While
   END

---

association solutions. Finally, through phase III, constructing the feasible solution set based on the forgoing two phases and selecting out the optimal user association solution from the solution set.

Before describing the algorithm, we first define the parameter function $f_\lambda(\mathbf{T})$,

$$\min f_\lambda(\mathbf{T}) = (\lambda \mathbf{P} + \Gamma)\mathbf{T}$$
$$s.t. \quad \mathbf{T} \in \mathcal{T} \tag{9.14}$$

where λ is the slope of the two points, which can also be regarded as the search direction. Assume $\forall \mathbf{T}_1 \in \mathcal{T}$ and $\forall \mathbf{T}_2 \in \mathcal{T}$ exist and we have $f^1 = (f_1^1, f_2^1)$, $f^2 = (f_1^2, f_2^2)$, then we have:

$$\lambda = \lambda(f^1, f^2) := \frac{(f_2^1 - f_2^2)}{(f_1^2 - f_1^1)} \tag{9.15}$$

The parameter function of Equation 9.13 can be solved by many algorithms such as genetic algorithm. Assume $\mathbf{T}^*$ be one solution of Equation 9.13, then $(\mathbf{PT}^*, \Gamma\mathbf{T}^*)$ is a supported non-dominated point. In the following, the detail of the algorithm is shown through the step 1, and the initial feasible solution $\mathbf{T^{UL}}$ and $\mathbf{T^{DR}}$ are gained. Based on the two solutions, supported non-dominated points f^{UL} and f^{DR} can be found. Assume $f^+ := f^{UL}$, $f^- := f^{DR}$, the search direction can be obtained with function in Equation 9.14 and $\lambda = \lambda(f^+, f^-)$. In step 5, we obtain point $f_\lambda^* = (f_1^*, f_2^*)$ and the optimal solution $\mathbf{T^*}$. If $f_\lambda^* < \lambda f_1^+ + f_2^+$, then the point is supported non-dominated. After repeating step 5, the supported non-dominated solution set $\mathcal{T}$ can be found.

The solutions calculated in phase I are the supported non-dominated points, which is only part of the solution space. However, the corresponding points of these solutions locate at the vertexes of the point space (as shown in Figure 9.2). The non-extreme points at the border of the point space and the non-supported non-dominated points can not be obtained. In phase II, we intend to find these points through searching in the triangle area.

$Delta(f^+, f^-)$. In addition, to reduce the complexity of the algorithm, when proposition 1 is satisfied, no unsupported and non-dominated points exist.

Proposition 9.1 *If two unsupported non-dominated points f^+ and f^- satisfy function in Equation 9.16, then none unsupported non-dominated points exist in the triangle area constructed by the two points*

$$f_1^- - f_1^+ = 1 \quad or \quad f_2^+ - f_2^- = 1 \tag{9.16}$$

When f^+ and f^- fail to meet Proposition 9.1, then phase II can be implemented.

In phase II, $KBest$ [16] is a sort algorithm, which can determine the Kbest solutions $\{\mathbf{T^1}, \mathbf{T^2}, \ldots, \mathbf{T^K}\}$ within the area of $\Delta(f^+, f^-)$, these solutions have the following relationship:

- $f_\lambda(\mathbf{T^i}) \leq f_\lambda(\mathbf{T^{i+1}}), i = 1, 2, \ldots, K - 1$

- for any unsupported non-dominated solution $\mathbf{T} \notin \{\mathbf{T^1}, \mathbf{T^2}, \ldots, \mathbf{T^K}\}$, which satisfies $f_\lambda(\mathbf{T^K}) \leq f_\lambda(\mathbf{T})$

The $KBest$ algorithm will return to the Kth solution and its corresponding unsupported non-dominated point, which have the minimum parameter value $f_\lambda(\mathbf{T})$.

In phase II, the triangle search area $\Delta(f^+, f^-)$ is constructed firstly. The vertexes of the triangle are (f_1^+, f_2^+), (f_1^-, f_2^-), and (f_1^-, f_2^+), respectively. The search direction is $\lambda(f^+, f^-)$. The order of parameter $f_\lambda((\mathbf{T}))$ can be obtained by *KBest* algorithm. When $f_\lambda((T))$ reaches the upper bound $UB := \lambda f_1^- + f_2^+$, the search stops, and meanwhile the optimal solution $\mathbf{T}^t$ and the corresponding parameter $f^t = f_\lambda(\mathbf{T}^t)$ is returned by method. Through substituting this parameter to function $NonDom(f^t)$, if satisfies, then f^t can be the supported non-dominated point. Then solution $\mathbf{T}^t$ can be included in the solution space $\mathcal{T}$, the upper bound can be updated by $UB = \mathbf{max}\{\lambda f_1^- + f_2,$ $\lambda f + f_2^+\}$.

All the feasible solutions for BiMMAP have been got in phase I and II. Here, we search the optimal solution for the sub-problem of OSP. Assume $g = \sum_{n \in S} \sum_{i \in U} T_{i,l}^n(t) 1/\gamma_{i,l}^n(\tau) = T\gamma$, phase III is shown in Algorithm 9.3.

In phase III, all the solutions selected in phase I and phase II are substituted to the function in Equation 9.9. After comparing the corresponding latency values by the bubble approach of step 3, the solution T^o is chosen as the optimal result for the user association problem.

Algorithm 9.3 Search the optimal solution

1: Initialization

2: Assume the number of the feasible solutions is K, the number of search time $t = 1$

$$\mathbf{T}^o = \mathbf{T}^1, g^o = \mathbf{T}^o \gamma^o$$

3: While $t \leq K$, do {

$\qquad t = t + 1,$

$\qquad g^t = \mathbf{T}^t \gamma^t$

$\qquad$ IF $g^t < g^o$, then

$\qquad\qquad g^o = g^t,$

$\qquad\qquad \mathbf{T}^o = \mathbf{T}^t$

$\qquad\qquad\qquad$ }

$\qquad$ END IF

$\qquad$ END While

4: Output the optimal solution $\mathbf{T}^o$.

$\qquad$ END

9.4 Simulation Results and Discussions

In this part, the effectiveness of the proposed user association scheme will be demonstrated by simulation. Within the simulation, we compare the performance of our scheme with several other schemes. The simulation parameters are given as follows.

One macrocell and 100 small cells are included in the C-RAN SCN scenario. The coverage area of the macrocell is 1000 m while that of small cell is 20 m. Other related simulation parameters are defined as: $v = 30\,km/h$, $c = 3 \times 10^8\,m/s$, $w = 1\,MHz$, $N_0 = -100\,dB$.

In Figure 9.3, the lines related to the network latency under the proposed TPSA and two other candidate association schemes are plotted. The candidate schemes are maximum SINR method (Max-SINR) with which users chooses the small cell with maximum SINR and QoS-based association algorithm (QOSA) that users access to the small cell that guaranteeing the minimum transmission rate. From the graph, it's obvious that with the TPSA, the network latency is lower than the other schemes, it can be concluded that users the proposed scheme is more adapt to the mobile network scenario. Moreover, the

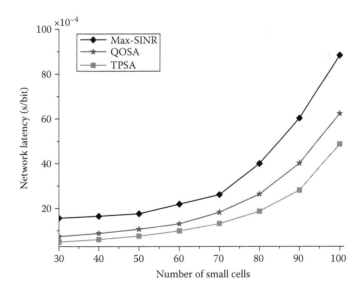

Figure 9.3 Network latency under the method of Max-SINR, QOSA, and the proposed TPSA.

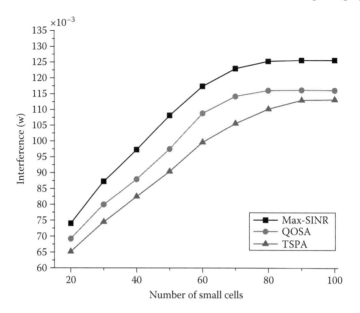

Figure 9.4 Interference suffered by the small cell under the method of Max-SINR, QOSA, and the proposed TPSA.

network latency has an increasing tend when the number of small cells becomes larger.

The results presented in Figure 9.4 show the comparison of Max-SINR, QOSA, and the proposed TPSA from the view of interference. From the graph, it can be obtained that as the number of small cells increase, the interference also becomes larger. The interference value increases until to an acme and then becomes steady. This happens because the maximum interference suffered by the macrocell is limited. When the interference generated from the small cells increases to a threshold, no more small cell are allowed to serve the users. Moreover, with our user association scheme, less interference is consumed.

The energy consumption of small cells is shown by the lines delineated in Figure 9.5. The schemes compared are the same with Figures 9.3 and 9.4 that are Max-SINR, QOSA, and the proposed TPSA. It can be obtained that with TPSA, small cell save more energy than other schemes. Moreover, the saved power shows a rising trend as more small cells enter the network.

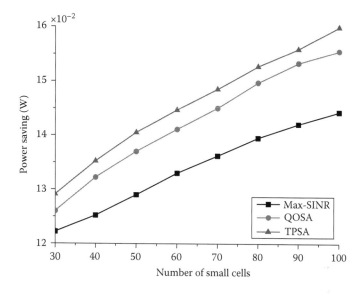

Figure 9.5   Energy consumption under the method of Max-SINR, QOSA, and the proposed TPSA.

9.5   Conclusions

In this chapter, to choose appropriate small cells for users, we studied the user association issue in a C-RAN SCN scenario. Combining with network latency, we constructed a user association optimization problem. What's more, considering the rising CO_2 emission and the necessary network throughput, energy saving and interference limitation are also incorporated in the optimization problem. To solve the user association optimization problem, a TPSA was proposed with the concept of Pareto Optimality. With this algorithm, appropriate small cells and physical resources are chosen for users while minimizing the network latency, minimizing the overall energy consumption and reducing the network interference. In addition, we considered the transmit latency for users. Since the algorithm computation time and software implementation time in the cloud are also crucial to the network latency.

References

1. Cisco. Cisco visual networking index: Global mobile data traffic forecast update, 2014–2019. White paper, Cisco, Technical Report, Feb. 2014.

2. C. Liu, K. Sundaresan et al. The case for re-configurable backhaul in cloud-RAN based small cell networks. *IEEE INFOCOM*, Turin, pp. 1124–1132, Apr. 2013.

3. H. Wen, P. K. Tiwary, and T. Le-Ngoc. Wireless virtualization. *Springer Briefs in Computer Science*, Sept. 2013.

4. C. Liang and F. R. Yu. Wireless network virtualization: A survey, some research issues and challenges. *IEEE Communications Surveys & Tutorials*, 17(1):358–380, Mar. 2015.

5. G. Tseliou, F. Adelantado, and C. Verikoukis. Resources negotiation for network virtualization in LTE-A networks. *IEEE ICC*, pp. 3142–3147, 2014.

6. M. Kalil, A. Shami, and Ye Yinghua. Wireless resources virtualization in LTE systems. *IEEE INFOCOM Workshops*, Toronto, Canada, pp. 363–368, 2014.

7. M. I. Kamel, L. Bao, and A. Girard. LTE wireless network virtualization: Dynamic slicing via flexible scheduling. *IEEE VTC Fall*, Vancouver, Canada, pp. 1–5, 2014.

8. Y. Mao, B. Long, et al. Opportunistic spectrum sharing for wireless virtualization. *IEEE WCNC*, Istanbul, Turkey, pp. 1803–1808, 2014.

9. D. Garlisi, F. Giuliano et al. Deploying virtual mac protocols over a shared access infrastructure using MAClets. *IEEE INFOCOM Workshops*, Turin, Italy, pp. 17–18, Apr. 2013.

10. F. Wen and U. C. Kozat. Stochastic game for wireless network virtualization. *IEEE/ACM Transactions on Networking*, 21(1):84–97, May. 2013.

11. M. Hoffmann and M. Stauferand. Network virtualization for future mobile networks: General architecture and applications. *IEEE ICC*, Kyoto, Japan, pp. 1–5, 2011.

12. A. Tzanakaki and, M. P. Anastasopoulos et al. Virtualization of heterogeneous wireless-optical network and it infrastructures in support of cloud and mobile cloud services. *IEEE Communications Magazine*, 51(8):155–161, 2013.

13. G. Fettweiss. A 5g wireless communication vision. *Microwave Journal*, 55(12):24–36, Dec. 12.

14. C. G. Kung, C. Liu et al. Architecture and applications of a versatile small-cell, multi-service cloud radio access network using radio-over-fiber technologies. *IEEE ICC*, Budapest, pp. 879–883, 2013.

15. L. Massoulie and J. Roberts. Bandwidth sharing: Objectives and algorithms. *IEEE/ACM Transactions on Networking*, 10(3): 320–328, 2002.

16. C. R. Pedersen, L. R. Nielsen, and K. A. Andersen. The bicriterion multimodal assignment problem: Introduction, analysis, and experimental results. *INFORMS Journal on Computing*, 20(3):400–411, 2008.

17. E.L. Ulungu and J. Teghem. The two phases method: An efficient procedure to solve bi-objective combinatorial optimization problems. *Foundations of Computing and Decision Sciences*, pp. 149–165, 1995.

Index

Printed and bound by CPI Group (UK) Ltd, Croydon, CR0 4YY

23/10/2024

01777673-0012